Geocentricity: The Debates

Scott Reeves et al.

Geocentricity: The Debates

ISBN: **1530517443**
ISBN-13: **978-1530517442**

Books by Scott Reeves:

The Big City
Demonspawn
Billy Barnaby's Twisted Christmas
The Dream of an Ancient God
The Last Legend
Inferno: Go to Hell
Scruffy Unleashed: A Novella
Colony
A Hijacked Life
The Dawkins Delusion
The Newer New Revelations
Death to Einstein!
Death to Einstein! 2
The House at the Center of the Worlds
The Miracle Brigade
Tales of Science Fiction
Tales of Fantasy
The Chronicles of Varuk: Book One
Soldiers of Infinity: a Novelette
Snowybrook Inn: Book Four
Welcome to Snowybrook Inn
Liberal vs. Conservative: A Novella
Zombie Galaxy: The Outbreak on Caldor
Apocalyptus Interruptus: A Novella
Temporogravitism and Other Speculations of a Crackpot
A Crackpot's Notebook, Volume 1

Graphic Novels:
The Adventures of Captain Bob in Outer Space

Geocentricity:
The Debates

Introduction

This book consists of exchanges I have had with people on YouTube over the past few months (it is March 2016 as I write this). The subject was geocentricity. In case you don't know, geocentricity is the theory that Earth is stationary at the center of the universe. Yes, believe it or not, in the modern world there are still people who actually advocate such a model of the universe. And as I've discovered, many people have an irrational hatred for anyone who dares to so much as utter the word "geocentricity." Only stupid, whacked out, scientifically illiterate, Bible-thumping, delusional, insane, anti-science, murderous, pedophilic, ignorant, conspiracy-theorists could possibly believe Earth is at the center of the universe. Despite the fact that, according to Einstein, from the viewpoint of an observer stationary relative to the Earth, it is perfectly valid to say that Earth is stationary at the center of the universe. But of course, if you ask most modern relativists, I just told you a bald-faced lie. Einstein would never have said such a thing.

But I didn't lie. I told you the bald-faced truth. It's a truth which most of Einstein's supporters do not like to admit. In some cases, they don't even seem to be aware that their own theory actually supports geocentricity, because it MUST.

But this brief introduction isn't the actual debate, so I'll save the arguments for later.

A couple of the usernames herein have been reduced to initials, to protect the innocent and all that.

I myself do not know who any of the persons I interacted with are in "real life." I don't know their credentials or the level of their scientific literacy. So I make no claim as to the veracity of anything they say. If you're uncertain of any of their claims, do your own research into what they're saying. The same goes for anything I say. Never accept anything at face value, no matter which side is saying it or how much authority they appear to have.

It should also be noted that, where possible, this book is being given away for free. I am not doing this book for money. The price of the paperback version is basically just the cost of manufacturing charged by the printer.

Also, the fact that my comments appeared in any given YouTube video's comments section does not mean that I in any way endorse or agree with what is being said in any specific video. It just means that I watched the video and perused its comments section, and therein found comments to which I felt like responding.

Hopefully the format of the debates is easy to follow. I use the label **"[someone] wrote:"** and then present what they wrote, exactly as it is posted on YouTube, without any editing, spelling or grammar correction. In most cases, the statement of the opponent is in quotes, followed by the response. Should be pretty self-evident once you start reading.

Also, occasionally there will be instances of +ScottReeves, or +CoolHardLogic, etc. This is the Google+ system, and simply identifies the user at whom the subsequent commentary is directed.

For anyone who is interested, I have numerous videos of my own on my YouTube channel that further expand on my thoughts on the subject of geocentricity and the pseudo-science that is Relativity

I want to stress that this book is by no means a comprehensive treatise on the subject of geocentricity, and I urge every reader to do an in-depth study into the subject and come to your own conclusions regarding it.

I want to thank everyone who took the time to enter the arena with me. I had fun doing this, and the opposition helped me to clarify my own stance on geocentricity. I think these were good debates on both sides.

Lastly I would like to congratulate the obvious victor: absolute Geocentricity.

Debate One

Scott Reeves vs. NGC 6205 and CoolHardLogic

Comments on YouTube video "Testing Geocentrism Part 2" by CoolHardLogic

NGC 6205 wrote:

+Scott Reeves

I agree with you that a geocentric reference frame is a valid reference frame. I use it when I observe the night sky. However, you do not understand that the absolute geocentric reference frame is debunked by demostrating that there are other reference frames which are also completely valid. That's because absolute geocentricism, or capital G geocentrism claims that it is the only, One True Reference Frame. That's what absolute means. It claims that geocentrism is more valid than heliocentrism, which is false. It is actually no more valid than marscentrism or venuscentrism or jupitercentrism. If you were to build a "Neo-Tychonic" model (which I suppose you adhere to) with Mars at its center, and then go to Mars, it would make exactly the same number of successful predictions as a geocentric Tychonic model used on Earth. Furthermore, you can take a simulation of the Neo-Tychonian model and let it run. Pause it. Go to the Sun and fix your position (your view of the simulation) above the Sun. Unpause the simulation. What you would see is pure heliocentrism. That means that the Neo-Tychonic model is actually a heliocentric model in which the observer is fixed relative to the Earth, rather than the Sun. It also goes vice-versa: heliocentrism is a Neo-Tychonic model in which the observer is fixed relative to the Sun rather than the Earth. Besides the position of the observer, the two models are completely equivalent. This is why absolute geocentrism is false: because there are many different valid reference frames besides the geocentric one, while the capital G geocentrism claims that it is the one true reference frame, more real or correct than other reference frames. Special and general relativity are not required to prove that absolute geocentrism is false. And this is why you are biased and a pseudoscientist. You adhere to one reference frame absolutely and reject the others, despite the fact that other reference frames are completely valid. Only reason I can find for this is religious in nature. If your reason for adhering to geocentrism is not religious, then please tell me which is? Why are you adhering to geocentrism as more true than heliocentrism or marscentrism or jupitercentrism?

Scott Reeves wrote:

+NGC 6205

"However, you do not understand that the absolute geocentric reference frame is debunked by demostrating that there are other reference frames which are also completely valid."

You are correct. I do not understand that the absolute Geocentric reference frame is debunked by demonstrating that there are other reference frames which are also completely valid. I do, however, understand that it is debunked by demonstrating that there are other reference frames which are also completely valid and equal to the absolute Geocentric reference frame. Which has not yet been done.

I completely understand the concept of reference frames, and I do not deny that every conceivable reference frame is a valid reference frame. But the simple existence of other valid reference frames does not mean that all valid reference frames are equal.

"That's because absolute geocentricism, or capital G geocentrism claims that it is the only, One True Reference Frame. That's what absolute means."

That may be your concept of what absolute G means, but to me it refers to absolute rest vs. relative motion, and actual center vs. relative center. Relativistic geocentrism says that there is no actual center to the universe and all motion in the universe is relative, with no absolute motion, period; while absolute Geocentrism says the universe has a center and Earth is stationary there, and all motion in the universe is relative to the frame of absolute rest as established by the absolutely motionless Earth. That's the distinction between absolute Geocentrism and relativistic geocentrism. Don't quote me on that, though; what's important here is the concept of absolute vs. relative motion, and an absolute center vs. a multitude of observer-dependent centers.

One thing to note here is that I believe there could be a center to the universe, establishing an absolute frame of rest, but without the Earth there. In which case, the universe would not be absolutely Geocentric. It would only be relativistically geocentric – relative to the center of the universe, regardless of whether some other planet was at the center or not. But unfortunately for relativity, there is no physical law that requires Earth to be at the center of its own observable universe, yet modern science both empirically and philosophically says that we are. More on that down below.

"If you were to build a 'Neo-Tychonic' model (which I suppose you adhere to) with Mars at its center, and then go to Mars, it would make exactly the same number of successful predictions as a geocentric Tychonic model used on Earth."

Yes, that is what relativity hypothesizes. Now let's all go to Mars and test it.

"This is why absolute geocentrism is false: because there are many different valid reference frames besides the geocentric one, while the capital G geocentrism claims that it is the one true reference frame, more real or correct than other reference frames."

Again, that's your concept of what capital G geocentrism claims. But the existence of many different valid reference frames does not make absolute Geocentrism false. Absolute Geocentrism (at least this absolute Geocentrist) does not deny the existence of other reference frames, nor does it claim that it is more real or correct than other reference frames. It only claims that it is absolute, and all other frames must be considered relative to it.

"Special and general relativity are not required to prove that absolute geocentrism is false."

Correct. They're required to prove their hypothesis that all reference frames are equal by putting it through the scientific method. They have not yet finished that task. Thus far, they have merely gathered evidence from within a geocentric reference frame, whether it be absolute or relativistic.

"And this is why you are biased and a pseudoscientist. You adhere to one reference frame absolutely and reject the others, despite the fact that other reference frames are completely valid."

As I said, I do not reject other reference frames as invalid. I merely say that the absolute Geocentric frame is unique among all frames.

"Only reason I can find for this is religious in nature. If your reason for adhering to geocentrism is not religious, then please tell me which is? Why are you adhering to geocentrism as more true than heliocentrism or marscentrism or jupitercentrism?"

Maybe I'm misunderstanding what you mean by "more true," but I can only say again that I regard the absolute Geocentric reference frame as unique and absolute among a multitude of reference frames.

As for why I adhere to absolute Geocentrism, for me, religion and

God have nothing to do with my reasons. Here is my reasoning:

Since modern mainstream science admits that we are empirically at the center of our observable universe, then it follows that if our observable universe is actually the ENTIRE universe, then Earth is most definitely motionless at the center of the actual universe, thereby establishing the Earth as an absolute reference frame. Thus, in order for the relativistic geocentric reference frame to be the correct choice when choosing between a relativistic geocentric or an absolute Geocentric frame, relativity must say that everyone is at the center of his/her/its own observable universe, including an observer on a planet at the edge of our observable universe, which observer must therefore be able to see something beyond the edge of our (observers on Earth) observable universe.

But anything outside our observable universe is...wait for it...unobservable, and so outside the scope of rational scientific inquiry by scientists on Earth. So before relativity can be claimed to have been properly tested by the scientific method, observers on Earth must travel to the edge of our observable universe to make observations confirming that there is a universe beyond Earth's observable universe, and that that point on the edge is the center of its own observable universe. Conveniently enough for relativity, such a test can never be performed, since according to current cosmological theory, Earthers can never reach the edge, because the edge of our observable universe is expanding away from us in all directions at faster than the speed of light. This makes relativity a pseudo-science, since its proponents wrongly tout it to the public as having been scientifically tested, even though it can never actually be scientifically tested.

Therefore as far as science as practiced by observers on Earth will ever be concerned, the observable universe is all that there is, and therefore we are absolutely at rest at the center of the entire scientifically observable universe.

Look at it this way. If relativists ever find an edge to the universe, that disproves their theory and proves absolute Geocentrism. So your theory cannot allow an edge to the universe, or at least can only allow an edge that is either unreachable by any observer in any reference frame, or is at an infinite distance from Earth, which amounts to the same thing. Therefore relativity can only ever prove absolute Geocentrism, and can never prove itself. To prove itself, relativity must prove that an edge does not exist. And how do you prove that something does not exist? Ask an atheist how he/she empirically proves that God does not exist.

You'll note that at no time in my argument do I appeal to God or the Bible to decide between absolute Geocentrism or relativistic geocentrism. So you are incorrect in your contention that I must be advocating absolute Geocentrism solely on religious grounds. Yes, I am a Christian, but whether we are or are not absolutely at the center of the universe makes no difference to my belief in God or to my ego. I choose the absolute Geocentric frame merely because I don't want to choose the pseudo-scientific alternative.

NGC 6205 wrote:

+Scott Reeves

I have to say that you make some good points and that your civilized debating style differs much from the debating style and arguments of the vast majority of people promoting, let's be honest, extremely fringe ideas. I commend that. However, I think you've misinterpreted the nature of science. You said: "Therefore relativity can only ever prove absolute Geocentrism, and can never prove itself. To prove itself, relativity must prove that an edge does not exist." Science is not in the job of proving theories. Science is in the job of constructing ever better models, which explain all the phenomena that previous models could explain plus new data previously unexplained. It must also make successful predictions. Simplicity is also valued, although it is not that important. The point is, if you have a successful theory which explains a lot and has never been disporved, you can allow yourself to derive assumptions from the model which you consider true unless disproved, although they have not been proved. For example, until 20 years ago we hadn't found a single planet orbiting a star other than the Sun. But scientists were long before that taking for granted that there are planets orbiting orbiting other stars. Why? Because they were pseudoscientists? Because they were dogmatic? Because they were atheists trying to shake our faith in a personal God? Of course not. They were doing that because that's a conclusion that you derive when you look at the bigger picture: The Sun is an ordinary star with planets, there are many other stars in the galaxy, we can see stars being born from large molecular clouds, we have a sensible theory of the formation of the Solar System, and so on and so on... Given this knowledge, scientists could say: "We can be 95% sure (or 70%, or 99%, it doesn't matter) that there are other planets in the galaxy." Science is all about probabilities. We cannot be 100% sure that there isn't something seriously wrong with the theory of plate tectonics, or quantum mechanics, or any other theory for that matter, but we behave like there isn't, because the probability of there being something wrong with the theories is very low. That's science. If you want absolute proof, go to mathematics or logic, because science cannot give you that. We have the same situation here: We know the fundamental forces of nature: gravity, electromagnetism, the weak nuclear force and the strong nuclear force. We know planets and stars are made of protons and neutrons and electrons, they interact via these forces, the Earth is also made of this stuff. We can see other planets orbiting their stars. They are not at the center of our observable universe by definition, why should the Earth be at the center of the entire universe? Why should the center of the observable universe as seen from the Earth be the center of the entire universe, but not the center of the observable universe as seen from these planets? Given that we know that the Earth is not special in the physical sense (because it's made of the same stuff and subjected to the same forces), there is no reason to believe that it just so happens that the Earth is at the center of the universe. Given the size of the universe, it is much, much more probable that the center of the universe happens to be in some other place, if there is any center at all. It's simply a matter of probabilities. It is not sensible or scientific to adhere to a very low probability model, although it has not been disproved. That's what science says: Geocentrism is simply not sensible or probable, given the big picture, which includes all our knowledge of physics, astronomy and cosmology. It is much more probable that the Earth, being made of the same stuff and subjected to the same forces as the other planets, is in the same situation as them: Not in the center of the universe and orbiting its star like the laws of physics dictate. Absolute proof isn't necessary, and I definitely don't lose sleep over it.

<u>Scott Reeves wrote:</u>

+NGC 6205

Thanks for your comments. I actually appreciate the opportunity to be challenged to defend what I'm saying.

"Science is not in the job of proving theories. Science is in the job of constructing ever better models, which explain all the phenomena that previous models could explain plus new data previously unexplained. It must also make successful predictions."

Science is, however, in the job of testing the predictions its theories make, and science has not yet done that with relativity. It has only gathered data from within a geocentric reference frame, which does not speak to the issue of whether the absolute Geocentric or the relativistic geocentric model is correct. The true test of relativity will be made when we leave whichever sort of geocentric reference frame we are currently in.

"We cannot be 100% sure that there isn't something seriously wrong with the theory of plate tectonics, or quantum mechanics, or any other theory for that matter, but we behave like there isn't, because the probability of there being something wrong with the theories is very low."

Actually, the probability of there being something wrong is NOT very low. We can be 100% sure that there is something seriously wrong with either quantum mechanics or relativity, since they are not compatible. The search for a theory of everything is an implicit admission that we are 100% certain that there is something seriously wrong with one or the other, or possibly both.

"They are not at the center of our observable universe by definition, why should the Earth be at the center of the entire universe?"

Because Earth is by definition and by empirical evidence at the center of our observable universe, and anything beyond our observable universe, the so-called "entire" universe, is unobservable and thus beyond the realm of science. Why should the Earth, against all empirical evidence, NOT be at the center of the entire scientifically-observable universe?

"Why should the center of the observable universe as seen from the Earth be the center of the entire universe, but not the center of the observable universe as seen from these planets?"

Because if our observable universe IS the entire universe, and we have no empirical evidence that it is not, then obviously a planet near the edge of the entire universe cannot possibly be at the center of its own observable universe. Why should a point on the circumference of a circle not be at the center?

"Given that we know that the Earth is not special in the physical sense (because it's made of the same stuff and subjected to the same forces)..."

Actually, if Earth is at the center of the entire universe, then it's NOT subject to the same forces as all the other stars and planets. It's the only planet in the universe that would feel the combined gravitational force of all the mass in the universe, equally from all directions.

"Given the size of the universe, it is much, much more probable that the center of the universe happens to be in some other place, if there is any center at all. It's simply a matter of probabilities."

You mean it's more probable that the center of the universe would be in some other place where there is no life similar to ours? Given the size of the solar system, it's much more probable that humans would have arisen on some other planet. Oh wait, it's not, because science recognizes that Earth occupies a special zone in the solar system. We're in the so-called Goldilocks Zone precisely because science recognizes that there is something special about this narrow strip of space around our sun. If the center of the universe happens to be a similar sort of Goldilocks Zone (maybe a motionless, non-rotating Earth is the most stable place in the universe, and thus somehow most conducive to life, I don't know), then it would actually be more probable that we would be at the center than elsewhere in our universe.

Your statement about probabilities is meaningless unless you're willing to claim that science knows all there is to know about how the universe works. If we're at the center against all probability, then it either means we're exceptionally lucky, or we don't have the proper information to accurately estimate the probabilities.

"It is not sensible or scientific to adhere to a very low probability model, although it has not been disproved."

Again, it's only a low-probability model if you arrogantly assume that you have a complete understanding of the universe. What your'e implicitly saying is that it's more sensible and scientific to choose the relativistic geocentric model over the absolute Geocentric model. But it's not sensible or scientific to choose the model that is part of a theory that has at least a 50% probability of ultimately being shown to be incorrect (quantum mechanics and relativity can't both be correct).

"That's what science says: Geocentrism is simply not sensible or probable, given the big picture, which includes all our knowledge of physics, astronomy and cosmology."

That's not what science says. That's what people who assume they have an accurate big picture say.

"It is much more probable that the Earth, being made of the same stuff and subjected to the same forces as the other planets, is in the same situation as them: Not in the center of the universe and orbiting its star like the laws of physics dictate."

The laws of physics dictate that a system of bodies orbits the center of mass of the entire system. Which is what our sun is doing: it's orbiting the center of mass of the entire universe. And Earth happens to be located at that center. That's why Earth is the only place in the universe where you'll see a star appearing to orbit one of its smaller planets.

CoolHardLogic wrote:

+Scott Reeves

"Science is, however, in the job of testing the predictions its theories make, and science has not yet done that with relativity. It has only gathered data from within a geocentric reference frame,."

False. Measurements validating GR are not confined to Earth.

"The true test of relativity will be made when we leave whichever sort of geocentric reference frame we are currently in."

No. And you have no a priori reason to assume a "geocentric reference frame". You're just being intellectually dishonest.

"We can be 100% sure that there is something seriously wrong with either quantum mechanics or relativity, since they are not compatible."

False dichotomy. Both are extensively tested and work within the domains to which they apply. Germ Theory is incompatible with GR.

The inability to reconcile them into a single relativistic germ theory doesn't mean that one is wrong. Again you're being intellectually dishonest.

"The search for a theory of everything is an implicit admission that we are 100% certain that there is something seriously wrong with one or the other, or possibly both. "

No, it isn't. Your conclusion is a non-sequitur borne of your desire for your conclusion to be correct.

"Because Earth is by definition and by empirical evidence at the center of our observable universe,..."

By definition? Regardless, you appear to utterly fail to understand the concept of a light horizon, and the simple fact that any observer anywhere in the Universe will appear to be at its centre.

"Because if our observable universe IS the entire universe, and we have no empirical evidence that it is not, then obviously a planet near the edge of the entire universe cannot possibly be at the center of its own observable universe. "

Argument from ignorance and special pleading. But thanks for demonstrating that you don't understand light and that you appear to think that the Universe is a fixed size bubble with an edge. That's just priceless.

"Why should a point on the circumference of a circle not be at the center? "

Because the universe is neither a circle nor a sphere with an edge.

"Actually, if Earth is at the center of the entire universe, "

And sinc eyou have no reason to assume the former, everything you said after this was completely pointless.

"then it's NOT subject to the same forces as all the other stars and planets."

Special pleading.

"It's the only planet in the universe that would feel the combined gravitational force of all the mass in the universe, equally from all directions."

Assuming your conclusion as your premise. Bad idea.

"We're in the so-called Goldilocks Zone precisely because science recognizes that there is something special about this narrow strip of space around our sun."

And so now you want to co-opt science when it suits you. How disingenuous. Also, the strip to which you refer isn't as narrow as you would like to imply to sate your apparent need to feel special.

"If the center of the universe happens to be a similar sort of Goldilocks Zone "

Again, you have no reason to even suppose that as a premise other than your desire to engage in circular reasoning.

"then it would actually be more probable that we would be at the center than elsewhere in our universe."

No, that conclusion doesn't follow from your contrived premise in any way, shape or form.

"Your statement about probabilities is meaningless unless you're willing to claim that science knows all there is to know about how the universe works."

Non-sequitur conclusion again. And hypocrisy, given that you were just making bullshit claims about probabilities.

"Again, it's only a low-probability model if you arrogantly assume that you have a complete understanding of the universe."

Straw man. And hypocrisy again given your obvious and rather desperate attempts to want to be at the centre of the Universe.

"What your'e implicitly saying is that it's more sensible and scientific to choose the relativistic geocentric model over the absolute Geocentric model."

Even if there were a relativistic geocentric model, it would still be bollocks.

"But it's not sensible or scientific to choose the model that is part of a theory that has at least a 50% probability of ultimately being shown to be incorrect (quantum mechanics and relativity can't both be correct)."

Repeating this claim again huh? It had no bearing on anything when you first mentioned it, and it still has no relevance now. So NGC 6205 pointed out that "Geocentrism is simply not sensible or probable, given the big picture, which includes all our knowledge of physics, astronomy and cosmology."

And you hilariously respond with: "That's not what science says. That's what people who assume they have an accurate big picture say."

OH go on then, I'll briefly humour you and your failure to understand the information you've been supplied with: What does science say then? Don't forget to name the scientists concerned and reference papers supporting your claims.

"The laws of physics dictate that a system of bodies orbits the center of mass of the entire system. Which is what our sun is doing: it's orbiting the center of mass of the entire universe."

Wrong again. You're really not very good at this are you? "

"And Earth happens to be located at that center."

HAHAHAHAHAHAHAA!! You think Earth is "the centre of mass of the entire Universe". You don't have a fucking clue about physics and now would be a good time for you to stop pretending that you do.

"That's why Earth is the only place in the universe where you'll see a star appearing to orbit one of its smaller planets."

HAHAHAHAHAHAHAHA!! There you go with your special pleading again. Instead of engaging in this tedious, fallacious bullshit that you seem to be so very keen on, why don't you provide a paper showing conclusively that Earth is the centre of MASS of the entire Universe and that the Sun orbits it.

Scott Reeves wrote:

+CoolHardLogic

"False. Measurements validating GR are not confined to Earth."

Allowing for the sake of argument that you are correct, GR is an incomplete theory, and GR is not the sum total of Einstein's relativity. The part I object to is that all reference frames are physically equivalent. That is the heart of Einstein's theory, and it has yet to be tested by the scientific method.

"No. And you have no a priori reason to assume a "geocentric reference frame". You're just being intellectually dishonest."

Actually I do have an a priori reason. We are currently in a geocentric reference frame, so I'm assuming that frame because I am an observer within that frame. Since relativity puts forth the hypothesis that all reference frames are physically equivalent, relativity must test that hypothesis by gathering evidence from within other reference frames, not just a geocentric frame. I'm not being intellectually dishonest, you are.

"False dichotomy. Both are extensively tested and work within the domains to which they apply. Germ Theory is incompatible with GR. The inability to reconcile them into a single relativistic germ theory doesn't mean that one is wrong. Again you're being intellectually dishonest."

Then you tell me what a physicist such as Brian Greene means when he writes, "As they are currently formulated, general relativity and quantum mechanics cannot both be right" (The Elegant Universe, pg 3). Saying that one of them cannot be right does not mean that one of them is wrong? What is this, Orwellian doublethink?

What does Stephen Hawking mean when he writes, "We have general relativity, the partial theory of gravity..." The Theory of Everything, pg 112). Tell me by what screwed-up logic an incomplete theory of gravity does not equal an incorrect theory of gravity? If a theory is only half right, then it is wrong, period. Almost right doesn't count.

But now I suppose I'm going to be ridiculed as a quote miner for citing authoritative figures to back up what I said. Or let me guess. I'm taking them out of context.

And I'm not being intellectually dishonest. You are.

"By definition? Regardless, you appear to utterly fail to understand the concept of a light horizon, and the simple fact that any observer anywhere in the Universe will appear to be at its centre."

The commenter to whom I was responding said that we were 'by definition' (his words) at the center of our observable universe. I was merely agreeing with his definition. As for the alleged appearance that I utterly fail to understand the concept of a light horizon, appearances can be deceiving. As for the alleged 'simple fact' that any observer anywhere in the universe will appear to be at its center, you're assuming that there is something beyond our own light horizon. In fact, you are making a hypothesis. Prove that an observer at our light horizon will see something beyond our horizon. Obey the rules of science. Go ye forth and test your hypothesis. Refusing to consider the possibility that there might be nothing beyond our light horizon does not make it true that there is something beyond our light horizon.

"Argument from ignorance and special pleading. But thanks for demonstrating that you don't understand light and that you appear to think that the Universe is a fixed size bubble with an edge. That's just priceless."

How do I know that you understand light enough to tell me that I don't understand light? Anyway, I never said that the universe is a fixed-size bubble. Are you denying that our observable universe has an edge? That's just priceless. We can only see about 14 billion light years out, and can see nothing past that. Seems like an edge to me. If you don't call that an edge, what DO you call it? Our light horizon, perhaps? A rose by any other name... Prove that there is something beyond our horizon. Or the edge of the universe, as I call it.

"Because the universe is neither a circle nor a sphere with an edge."

Special pleading.

"And sinc eyou have no reason to assume the former [that Earth is at the center of the entire universe], everything you said after this was completely pointless."

I do have reason to assume Earth is at the center of the entire universe: we have no empirical, scientific evidence of a universe beyond the edge of our observable universe (or beyond our light horizon, as you apparently insist that it be called). It is YOU who have no valid reason other than special pleading to assume that there IS something beyond our light horizon.

"Assuming your conclusion as your premise. Bad idea."

Actually, no. The premise is that Earth is at absolute rest at the center of the entire universe. The conclusion based upon the premise is what I said: 'It's the only planet in the universe that would feel the combined gravitational force of all the mass in the universe, equally from all directions.' Which is what I said in response to a non-geocentric statement that Earth is subject to the same forces as all the other matter in the universe.

"And so now you want to co-opt science when it suits you. How disingenuous."

I'm not co-opting science. I've been using science all along in my argument. I mentioned the concept of the Goldilocks Zone as a possible explanation of why we find ourselves at the center of the universe.

"Also, the strip to which you refer isn't as narrow as you would like to imply to sate your apparent need to feel special."

Actually it is fairly narrow compared to the overall size of the solar system. You should look it up sometime. As for my apparent need to feel special, I have no such need. Why do you presume to know what my needs are?

"Again, you have no reason to even suppose that as a premise other than your desire to engage in circular reasoning."

Why do you keep presuming to know what my needs and desires are? Can you tell me what my thoughts are as well? Quick, what am I thinking right now? Anyway, I guess my alleged desire is sort of like your desire to engage in similar circular reasoning when you insist without any evidence that every point in the universe is at the center of its own observable universe.

"No, that conclusion doesn't follow from your contrived premise in any way, shape or form."

The allegedly contrived premise to which you refer is that Earth is in a unique Goldilocks Zone at the center of the universe. The conclusion is that it would therefore actually be more probable that human life would evolve at the center than elsewhere in our universe.

If you believe my conclusion doesn't follow from my premise, then you must also believe the premise that Earth is in a Goldilocks Zone in our solar system doesn't lead to the conclusion that it's therefore more probable that human life would evolve on Earth than elsewhere in the solar system.

"Even if there were a relativistic geocentric model, it would still be bollocks."

IF there were? There has to be one. ALL relativists are implicitly geocentrists, they're just geocentrists who don't believe that the geocentric frame is a preferred or unique frame as do the other sort of geocentrists. So if you think relativity's geocentric model is bollocks (can I use that word too, or have you trademarked it?), then you are an anti-relativist. Welcome to the club.

"Repeating this claim again huh? It had no bearing on anything when you first mentioned it, and it still has no relevance now."

It had every bit of relevance when I first mentioned it, and it still does now. The choice on offer is between a relativistic geocentric reference frame and an absolute Geocentric reference frame. Relativity is pseudoscience, therefore the only real choice is the absolute Geocentric reference frame.

"OH go on then, I'll briefly humour you and your failure to understand the information you've been supplied with: What does science say then?"

What does science say about what? The big picture to which NGC 6205 referred? Science doesn't say anything. It's merely a tool that people use (or misuse) to test their hypotheses, and then make statements regarding their interpretation of the test results. My view of the big picture given to us by science and reason says that we're absolutely at rest at the center of the entire universe. You apparently hold the view that the big picture given to us by science and reason says that we're only at the center of our own observable universe, and likewise every other observer is at the center of his/her/its own observable universe. One of us has an incorrect view of the big picture. Can you guess which one of us I believe it is? I'll bet you can, since you seem to be intimately familiar with my needs and desires.

"Wrong again."

What's wrong? That the laws of physics dictate that a system of bodies orbits the center of mass of the entire system, or that the sun is orbiting the center of mass of the entire universe? If you think the former is wrong, then you really should study physics a bit more, and if it's the latter, then I call special pleadings on you.

"You're really not very good at this are you?"

Actually, I'm actually VERY good at this.

"HAHAHAHAHAHAHAHAA!! You think Earth is "the centre of mass of the entire Universe". You don't have a fucking clue about physics and now would be a good time for you to stop pretending that you do."

Did I say that I thought Earth was the center of mass of the entire universe? Show me where I said I thought Earth was the center of mass of the entire universe.

"HAHAHAHAHAHAHAHA!! There you go with your special pleading again. Instead of engaging in this tedious, fallacious bullshit that you seem to be so very keen on, why don't you provide a paper showing conclusively that Earth is the centre of MASS of the entire Universe and that the Sun orbits it."

I don't need to provide a paper showing conclusively that Earth is the center of MASS of the entire universe, because I never made that claim. Re-read my comments and then show me where I made that claim.

Debate Two

Scott Reeves vs. Nope

Comments on YouTube video "Testing Geocentrism Part 2" by CoolHardLogic

Scott Reeves wrote:

Einstein and Stephen Hawking, and many other reputable scientists, say geocentrism works, but we're supposed to believe you that it doesn't?

Nope wrote:

+Scott Reeves

No, you're supposed to believe the evidence gathered through observations, which clearly points out that geocentrism is bollocks. Also, I'm pretty sure none of those fellas actually believed for a second that geocentrism is an even remotely plausible explanation, but who knows? Even Newton said hillariously stupid shit back then so yeah.

Scott Reeves wrote:

+Nope

Here is the ColdHardTruth that all you relativity supporters are going to have to accept if you want to be relativists: there is absolutely not a single shred of evidence you can put forward to disprove the geocentric reference frame. The geocentric reference frame is a perfectly valid reference frame in relativity.

The only thing you can do is try to talk me and other Geocentrists out of advocating an absolute Geocentric reference frame. How do you do this? You point out that relativity forbids an absolute reference frame, and then try to convince us that relativity has a hundred years of empirical evidence behind it. You tell us that it is the most well-tested theory in the history of science, and only a fool would reject an argument with that magnitude of empirical weight behind it. And if those tactics don't work, you begin to pile on the mockery and the public humiliation in attempt to silence us.

That is ALL you can do. You all keep trying to prove that the Earth is in motion and the geocentric reference frame is invalid. News flash: that is most definitely not going to happen. If you think it will, you do not understand relativity. Trying to disprove the geocentric reference frame by saying that the Earth is definitely in motion is just the opposite of what I do by advocating the absolute Geocentric frame: you are trying to prove a frame of absolute motion. So by denying geocentrism in any form you're also working hard to try to disprove relativity.

When you eventually come to accept this ColdHardTruth and start putting forth the proper argument against absolute Geocentrism, I will still reject you, because I utterly reject relativity on the grounds that it is pseudo-science, so I really don't care if relativity forbids an absolute reference frame, and I reject the claim that relativity has the weight of empirical evidence behind it and is the most well-tested theory in modern science, because it cannot possibly have been properly tested, because all that alleged evidence is merely control data that has been gathered from within a geocentric frame and is awaiting replication in another reference frame at a cosmologically significant distance from Earth (yes, I know, run-on sentence).

Relativity is pseudo-science, and I don't care how many "reputable" and intelligent scientists support it, scientific and empirical truth is not decided by how many people believe in a theory.

<u>Nope wrote:</u>

+Scott Reeves

You know, you're saying all this nonsense right under a video that is part of a 12-episode series talking about why the geocentric view is inconsistent with our observations and therefore utterly useless for humanity. In every video there is at least 3 to 10 points why your model is unviable , many of them using simple math and basic scientific theories like gravity etc.

You know, the thing is, you can come up with as many frames as you want to describe the universe with the same amount of arrogance you display, but no matter how much you think reality depends on what science you "accept", at the end of the day it's always the practical usability that decides wich model is more plausible. If a model has efficient predicting capabilites, we use that and ignore the ones that hold no practical value to us or flat-out contradict our observations. We launch satelites into space and maintain them on a daily basis, we send probes to other planets, we predict the orbit of millions of celestial bodies, we look into the past of our universe... None of these things would be possible with the use of the geocentric model.

It's the same reason why there's a scientific concensus regarding general relativity: not because most scientist just happened to agree on it because they felt like it, but because it's a useful model with predictive powers. Your own GPS depends on that very theory.

So if you're done playing the victim of the "big conspiracy", please start addressing these videos point by point and refute them using evidence. Not this "well, my reference frame should be as valid as yours and relativitiy is pseudo-scientific anyway" crap, because that's nothing but whining. SHOW IT! Cheers.

<u>Scott Reeves wrote:</u>

+Nope.

"You know, you're saying all this nonsense right under a video that is part of a 12-episode series talking about why the geocentric view is inconsistent with our observations and therefore utterly useless for humanity."

You're kidding. I did not realize that. And absolutely nothing I said was nonsense. I stand by every word of it.

"In every video there is at least 3 to 10 points why your model is inviable, many of them using simple math and basic scientific theories like gravity etc."

So what you're saying is that in each video there are at least 3 to 10 points explaining why relativity is an inviable theory. And anyway, the Geocentric model which CHL attacks, in his first video at least, is an obsolete Geocentric model that no modern Geocentrist endorses, except perhaps for Fernieboy100, which makes him the perfect straw man for CHL .

"If a model has efficient predicting capabilites, we use that and ignore the ones that hold no practical value to us or flat-out contradict our observations."

Unfortunately for relativity, the predictions it allegedly makes are predictions for what you would expect to see from within a geocentric reference frame, which is precisely the frame from which we make those observations. This is why relativists, to strictly adhere to the scientific method, must perform those same observations from a cosmologically significant distance from Earth, and in a statistically significant number of reference frames, and replicate the results they obtained on Earth, from within the geocentric reference frame. To date, this has not been done, therefore relativity has not yet passed through the scientific method.

"We launch satelites into space and maintain them on a daily basis, we send probes to other planets, we predict the orbit of millions of celestial bodies, we look into the past of our universe... None of these things would be possible with the use of the geocentric model."

Yes, they would be possible with the geocentric model. They have to be possible with the geocentric model, or relativity is invalid. As for looking into the past of our universe: how would this in particular not be possible from an absolute Geocentric reference frame? Show me a Geocentrist who denies that light takes time to travel from distant parts of the universe, leaving aside any arguments as to the actual size of the universe?

"It's the same reason why there's a scientific concensus regarding general relativity: not because most scientist just happened to agree on it because they felt like it, but because it's a useful model with predictive powers."

I'll simply copy and paste an earlier answer: Unfortunately for relativity, the predictions it allegedly makes are predictions for what you would expect to see from within a geocentric reference frame, which is precisely the frame from which we make those observations. This is why relativists, to strictly adhere to the scientific method, must perform those same observations from a cosmologically significant distance from Earth, and in a statistically significant number of reference frames, and replicate the results they obtained on Earth, from within the geocentric reference frame. To date, this has not been done, therefore relativity has not yet passed through the scientific method.

"So if you're done playing the victim of the 'big conspiracy'"

I didn't mention any big conspiracy, and I didn't imply any big conspiracy. And I would in no way view myself as a victim if there were any such big conspiracy.

"please start addressing these videos point by point and refute them using evidence."

Why would I want to address point by point the videos of someone who is attempting to disprove relativity, whether or not CHL realizes that is what he is actually attempting? More power to him. I just recommend that he be a little less obnoxious in his attempts, and that he should attack a CURRENT Geocentric model, one propounded by most modern Geocentrists, at least in his first video.

"Not this 'well, my reference frame should be as valid as yours and relativitiy is pseudo-scientific anyway' crap, because that's nothing but whining. SHOW IT! Cheers."

I can only assume you mean I'm 'whining' that 'my' absolute Geocentric frame should be as valid as your relativistic geocentric frame. My absolute frame is actually valid and your relativistic one isn't, so I'm not whining. I'm just spreading truth.

Anyway, if that assumption is correct, then yes, the choice available to us is between an absolute Geocentric reference frame and a relativistic geocentric reference frame.

To date, NO observation has been put forward that cannot be accommodated by an absolute Geocentric reference frame. You can cite any sort of evidence you'd like: GPS, Focault's pendulum, geosynchronous satellites, geostationary satellites, parallax, aberration, particle accelerators, Hafele-Keating, cosmic ray muons - anything you'd like, and none of it contradictory to absolute Geocentrism. And according to relativists who actually understand relativity, all the observations support a relativistic geocentric reference frame as well. As I said earlier, you should be admonishing me for advocating an absolute Geocentric frame, not trying to disprove any sort of geocentrism whatsoever.

You have to find a way to choose between the two possible geocentric frames. Since the relativistic geocentric reference frame is dependent upon the truth of the Copernican principle, which is actually a hypothesis that has no empirical support and is basically a twin of the principle of relativity, then the choice is clear: relativity has not been properly tested, yet is presented to the public as scientific fact, making relativity a pseudo-science. Therefore, the choice is between absolute Geocentrism or a pseudo-science. Which does a true scientist choose?

Nope wrote:

+Scott Reeves

OK, so to cut this short, let's start this over again with you describing this so-called "absolute Geocentric model" of yours. And please, don't just refer me to your video. Maybe it's just that I'm an idiot, but I couldn't make any sense of it. I'm more of the reading type.

Scott Reeves wrote:

+Nope

I don't have any video that I recall which describes the modern geocentric reference frame in detail. Maybe the best way to describe it would be "The relativistic geocentric reference frame, except the absolute version of it." And there's really no difference between the two, other than the assertion that the absolute Geocentric reference frame is...well...absolute as opposed to relative. And it has a capital-G to distinguish it, following the convention of Phil Plaitt, the so-called Bad Astronomer. It looks a lot like the one CHL attempts to debunk in video 10, which he dubs Brahe 2.0. You'll note that he(?) doesn't claim that the model doesn't work or can't be modified to work, only that it fails a shave with Occam's Razor and is "Bollocks!" which doesn't speak to whether or not the model is viable. Occam's Razor isn't a physical law that governs the universe. If it violates Occam's razor, I don't care.

Not being sarcastic, but which video of mine did you watch and couldn't understand (all of them is not a helpful answer) because I've got like 150 or so on YouTube, and if you're going to start asking questions about one, I'll need to know which one you're talking about.

Debate Three

Scott Reeves vs. EmperorZelos

Comments on YouTube video "Heliocentric Vs Concave Geocentric Model Of Our Universe" by Godrules

Scott Reeves wrote:

+Mike Vizioz

"There is absolutely no reason to believe in something like that unless you are out to prove the validity of the bible."

How about a desire to be scientifically accurate? The choice is either absolute Geocentrism or relativity. Since relativity requires the existence of an unscientific concept, namely the unobservable universe, to be valid, relativity is not a scientific theory, yet is presented as scientific fact. Therefore the choice is either absolute Geocentrism or pseudoscience.

From a religious and Biblical perspective (yes, I am a Christian) I couldn't care less whether we're at the center of the universe or not. Makes no difference to my belief in God. I only want to follow the scientifically sound view of the universe, and protest all you want, that view is Geocentrism.

Scott Reeves wrote:

+EmperorZelos

Prove it. Have you gone to other points in the universe to check your assertion?

EmperorZelos wrote:

+Scott Reeves

EVERY point of the universe will be seen as the center you idiot and big bang/ relativity does nto depend upon a larger universe. Go back to highschool

Scott Reeves wrote:

+EmperorZelos

Prove that every point in the universe will be seen as the center. Get in your little spaceship and fly a few thousand light years away from Earth and start proving the Copernican principle.

And Big Bang/relativity DOES depend upon a larger universe. If we're at the center of the observable universe, which science acknowledges that we are, then in order for us not to be motionless at the center of an absolute Geocentric universe, there MUST be a larger universe beyond the observable universe. YOU go back to high school.

Emperor Zelos wrote:

+Scott Reeves

"Prove that every point in the universe will be seen as the center. Get in your little spaceship and fly a few thousand light years away from Earth and start proving the Copernican principle. "

It's demonstrable by the laws of physics, they are frame independed which means all points are equal.

"And Big Bang/relativity DOES depend upon a larger universe"

It doesn't, it might be a conclusion but it's not a dependence.

"If we're at the center of the observable universe, which science acknowledges that we are"

It doesn't, cite a single peer reviewed article from a reputable journal that says we are the center.

"then in order for us not to be motionless at the center of an absolute Geocentric universe, there MUST be a larger universe beyond the observable universe. YOU go back to high school. "

That's a non-sequitor.

The ceocentric model is dead and easily done so by simple newtonian physics because the earth moves around the sun, which moves around the center of the galaxy. We've measured these speeds and much else.

Scott Reeves wrote:

+EmperorZelos

"It's demonstrable by the laws of physics, they are frame independed which means all points are equal."

Which laws of physics, specifically?

"It doesn't, it might be a conclusion but it's not a dependence."

It is a dependence, because if the observable universe is all that there is to the universe, which obviously there can be no proof that there IS more to it given the inherently unobservable nature of that alleged more, and we're at the center of the observable, then we're absolutely, non-relatively at the center. Therefore relativity at least, and probably the Big Bang, depend upon there being more to the universe than what can be observed.

"The ceocentric model is dead and easily done so by simple newtonian physics because the earth moves around the sun, which moves around the center of the galaxy."

Prove that the Earth moves around the sun. All you can prove is that there is relative motion between the Earth and the sun. That is ALL you can prove.

Scott Reeves wrote:

+EmperorZelos

Forgot this part: "It doesn't, cite a single peer reviewed article from a reputable journal that says we are the center."

I'm not looking through peer-reviewed journals, wading through their tech-talk to find an instance where they say we're at the center of the observable universe. I can, however, throw all sorts of websites at you with quotes from physics professors and such saying that we are at the center of the observable universe. I'm not going to waste my time doing that, though, because I've got all that in one of my videos here on YouTube.

And anyway, how can you claim in one breath that every point in the universe will be seen as the center, and then deny in the next that science acknowledges that we're at the center of the observable universe? What is that, some sort of retard double-think to try to confuse the issue?

Emperor Zelos wrote:

"I'm not looking through peer-reviewed journals, wading through their tech-talk to find an instance where they say we're at the center of the observable universe. "

That's what you gotta do because webshites are worthless.

"I can, however, throw all sorts of websites at you with quotes from physics professors and such saying that we are at the center of the observable universe."

Not relevant unless the entire context of the quote is supplied.

"And anyway, how can you claim in one breath that every point in the universe will be seen as the center, and then deny in the next that science acknowledges that we're at the center of the observable universe? What is that, some sort of retard double-think to try to confuse the issue?"

First of you have not given any evidence that science says we are at the center, which I know is not the case because scientists know that all the laws of physics are frame invariant. This means no frame is prefered and special and all give equal predictions and as such tehre cannot be any centre without every point being the centre and if all points are the centre, do we really have a centre? No, it's just a product of the frame then.

<u>Scott Reeves wrote:</u>

"First of you have not given any evidence that science says we are at the center, which I know is not the case because scientists know that all the laws of physics are frame invariant."

You yourself gave the evidence in an earlier comment. I quote you from one of your comments on this very video: "+Godrules Idiot, they know exacly how to deal with the data because....guess what?

EVERY FUCKING POINT IN THE UNIVERSE WILL SEE ITSELF
AS BEING THE CENTER!?"

So unless you don't really know what you're talking about and thus everything you say to me is complete shite, we are at the center of the observable universe, and you have acknowledged it. The observer in the phrase "the observable universe" is us. When scientists talk about the observable universe, they're not talking about the observable universe as observed by Spock over on Vulcan.

That we are at the center of the observable universe is such a basic tenet of modern science that I don't NEED to slog through a bunch of peer-reviewed, "scientific" journals to search for a "credible" statement in support of my claim regarding modern science's position on the subject, any more than you have to do the same in support of your previous "EVERY FUCKING POINT IN THE UNIVERSE WILL SEE ITSELF AS BEING THE CENTER!?" comment.

What YOU need to do is slog through the scientific journals in search of empirical PROOF for your contention that EVERY point in the universe, not just ours, will see itself as the center. Guess what? THERE IS NO EMPRICAL PROOF OF THAT! Because the way you empirically prove it is to travel a cosmologically significant distance from Earth, perform your observations, and get the same results that you got on Earth. Unless you are aware of some secret space program that has done just that, then the claim that every point will see itself as the center of its own observable universe, which is a paraphrase of the Copernican principle, is an untested hypothesis.

"...which I know is not the case because scientists know that all the laws of physics are frame invariant. This means no frame is prefered and special and all give equal predictions and as such tehre cannot be any centre without every point being the centre and if all points are the centre, do we really have a centre? No, it's just a product of the frame then.?"

So you WERE talking about the principle of relativity earlier. The relativity principle is actually a hypothesis that says basically, as you put it, "...no frame is pefered [sic] and special and all give equal predictions..."

It's a HYPOTHESIS, closely tied to the Copernican principle (actually, another hypothesis), and it is exactly this hypothesis that makes relativity pseudo-science, because the hypothesis implicitly requires the existence of a larger universe beyond our observable universe. As I said, if the (our) observable universe is all that there is to the universe, AND we are at the center of it ("EVERY FUCKING POINT IN THE UNIVERSE WILL SEE ITSELF AS BEING THE CENTER!?"), then both the principle of relativity (read hypothesis) and the Copernican principle (read hypothesis) will fail. Thus, relativity and its adherents MUST prove that there is a larger universe beyond our observable universe, which, because it is implicitly unobservable and thus beyond the scope of rational scientific inquiry, is an impossible task for relativity.

So your statement "...no frame is prefered and special and all give equal predictions and as such tehre cannot be any centre without every point being the centre..." is an as-yet-untested hypothesis, a hypothesis that depends upon the existence of something that cannot be observed, making it a hypothesis that cannot be empirically tested. And since relativity is presented to the public as scientific fact, relativity is PSEUDO-SCIENCE.

So we're left with two options: 1) We are absolutely, non-relativistically at the center of the observable universe, and the observable universe is all that there is, or 2) We are relativistically at the center of the observable universe, but that's okay, because "EVERY FUCKING POINT IN THE UNIVERSE WILL SEE ITSELF AS BEING THE CENTER!?"

Option 1 is scientifically sound and is supported by ALL empirical evidence. Option 2 is scientifically unsound as it depends upon the existence of something that is as real as fairies and unicorns and the flying spaghetti monster.

Which option should a true scientist choose?

EmperorZelos has no clothes.

EmperorZelos wrote:

"You yourself gave the evidence in an earlier comment. I quote you from one of your comments on this very video: "+Godrules Idiot, they know exacly how to deal with the data because....guess what? EVERY FUCKING POINT IN THE UNIVERSE WILL SEE ITSELF AS BEING THE CENTER!?"

If every point is the center than no point is the center you imbecile.

"because the hypothesis implicitly requires the existence of a larger universe beyond our observable universe. "

It doesn't.

"So your statement "...no frame is prefered and special and all give equal predictions and as such tehre cannot be any centre without every point being the centre..." is an as-yet-untested hypothesis, a hypothesis that depends upon the existence of something that cannot be observed, making it a hypothesis that cannot be empirically tested."

It can and has been tested Scientists go with evidence you imbecile and geocentrism is not it. We know that earth is moving through space, our galaxy is moving and all

Scott Reeves wrote:

"If every point is the center than no point is the center you imbecile."

That's a mighty big if with no empirical evidence behind it. Every point is not the center, that's the point. There's only one point that is the center, and that point is us.

"It doesn't."

Does too.

"It [principle of relativity] can and has been tested"

How has it been tested? Particle accelerators? GPS? Astronomical observations? Hafele-Keating? Cosmic ray muons? Those aren't tests of the principle of relativity. They're merely the gathering of observations from within an Earth-based reference frame, observations which say that strange things happen when you move relative to the Earth, which doesn't contradict absolute, non-relativistic Geocentrism. Now, to properly test the principle following the scientific method, the same observations must be made, and the same results obtained, at a cosmologically significant distance from Earth. To date, that has not been done. "

"Scientists go with evidence you imbecile and geocentrism is not it."

You're correct. Scientists do go with evidence. Which is why relativists and anti-geocentrists do not qualify as scientists.

"We know that earth is moving through space, our galaxy is moving and all"

How do you know this? Cite me any empirical evidence that says it's Earth that is definitely moving, rather than just relative motion between Earth and something else. You cannot cite me any such evidence, because there IS none! I, however, can cite evidence that Earth is motionless: interferomter experiments.

I quote Albert Einstein: "For example, strictly speaking one cannot say that the Earth moves in an ellipse around the Sun, because that statement presupposes a coordinate system in which the Sun is at rest..."

"You fool!" he said with a flourish of his cape, "It is not I who am the imbecile! It is you!"

EmperorZelos wrote:

"That's a mighty big if with no empirical evidence behind it. Every point is not the center, that's the point. There's only one point that is the center, and that point is us. "

It isn't because all laws of physics have been tested and they are frame independed.

"They're merely the gathering of observations from within an Earth-based reference frame, observations which say that strange things happen when you move relative to the Earth, which doesn't contradict absolute, non-relativistic Geocentrism."

Geocentrism is dead since long ago so drop that already. We have gathered data in space, from other planets, satelites, as earth moves around the sun, sun through the galaxy and more.

"You're correct. Scientists do go with evidence. Which is why relativists and anti-geocentrists do not qualify as scientists."

Realitivity is confirmed to all levels on non-quantum levels and geocentrism is long since dead as it matches no fucking data.

"How do you know this? Cite me any empirical evidence that says it's Earth that is definitely moving, rather than just relative motion between Earth and something else"

Diurnal aberation, annual aberation, paralax, the dipole in microwave background radiation just to mention a small fraction of all evidence.

"You cannot cite me any such evidence, because there IS none! I, however, can cite evidence that Earth is motionless: interferomter experiments. "

Yet I did you imbecile.

Scott Reeves wrote:

"It isn't because all laws of physics have been tested and they are frame independed [sic]."

The speed of light at least, from the Geocentric viewpoint, is not frame independent. So ALL the laws are frame independent only if relativity is true. Relativity still has not been tested as required by the scientific method. Such testing cannot have been tested as required by the scientific method until such tests are replicated at a cosmologically significant distance from Earth, since non-relativistic Geocentrism is true if relativity is false.

"Geocentrism is dead since long ago so drop that already. We have gathered data in space, from other planets, satelites [sic], as earth moves around the sun, sun through the galaxy and more."

I will not drop Geocentrism, ever. I will chase it round the Moons of Nibia and round the Antares Maelstrom and round Perdition's flames before I give it up! Unless of course you're just saying you want to end our little back and forth now, which is okay with me. Anyway, Geocentrism is only "dead since long ago" by a consensus of pseudo-scientists, not by any empirical evidence against it. We may have gathered data in space, etc, as you say, but it wasn't gathered at a cosmologically significant distance from Earth, so it says nothing against Geocentrism. And the data gathered from other planets relates only to the planet in question, such as surface temperature, soil and atmospheric composition, etc. Have these planetary probes been making detailed astronomical observations regarding distant parts of the universe? Not that I've been hearing. And even if they have, they are still not being performed at a cosmologically significant distance from Earth, which is a requirement to properly test both relativity and the Copernican principle.

"Realitivity [sic] is confirmed to all levels on non-quantum levels and geocentrism is long since dead as it matches no fucking data."

No it's not and yes it does.

"Diurnal aberation [sic], annual aberation [sic], paralax [sic], the dipole in microwave background radiation just to mention a small fraction of all evidence."

All of which are easily explainable within a Geocentric universe if you care to look into the matter.

Here's a bit from Stephen Hawking. He's pretty smart, even though he believes in a pseudo-science, so you should listen to him: "...for our observations of the heavens can be explained by assuming either the earth or the sun to be at rest. Despite its role in philosophical debates over the nature of our universe, the real advantage of the Copernican system is simply that the equations of motion are much simpler in the frame of reference in which the sun is at rest." - Stephen Hawking, The Grand Design, pages 41-42

All the experiments you cited - diurnal aberration, annual aberration, etc - are all observations of the heavens, and if Stephen Hawking says they can be explained by assuming the Earth is at rest, I believe him. Plus, I've looked into each of those on my own, and Geocentrism can definitely explain them. Geocentrism can even explain geostationary and geosynchronous satellites, just in case you're going to bring those up.

You do realize that if there are any sort of observations showing Earth is definitely in motion, then relativity is an invalid theory, don't you? In which case, non-relativistic Geocentrism is your only alternative. So by arguing that there are observations that definitely prove the Earth is in motion, you are shooting yourself in the foot. If the Earth is definitely in motion, you're once again stuck having to explain why we can't detect Earth's motion using interferometers, which will lead you once again to the relativity of all motion, which will once again lead to your alleged detection of the Earth's definite motion, which will again destroy relativity...

"Yet I did you imbecile."

No, you didn't, you imbecile (can't we talk like mature people and stop the name-calling? I get it already; you think I'm an imbecile. Chill out, dude.). All you cited me are observations which show the relative motion between Earth and the universe, not any observations which show that the Earth is definitely the object in motion.

MikeVizioz wrote:

+Scott Reeves

"Geocentrism is only 'dead since long ago' by a consensus of pseudo-scientists"

...You just called every single scientist in history a pseudo-scientist.

Scott Reeves wrote:

+Mike Vizioz

That's all right. If you're wrong, you're wrong, no matter who you are. But I'm sure there are a few scientists out there in history and the present who weren't covered by my sweeping statement.

But maybe that was a bit too harsh. I'll be gentler and modify it to "Geocentrism is only 'dead since long ago' by a consensus of scientists who were fooled by a pseudo-science."

EmperorZelos wrote:

+Scott Reeves

"Relativity still has not been tested as required by the scientific method."

This is completely false, it has been tasted over and over and over again and it has succeeded at every point. It is tasted everyday by cellphones.

"Anyway, Geocentrism is only 'dead since long ago' by a consensus of pseudo-scientists, not by any empirical evidence against it."

Yeah no, it's dead by scientists because all evidence ever gathered goes against it.

"All of which are easily explainable within a Geocentric universe if you care to look into the matter."

Easily explainable? Look up cool hard logics video and see what explination is required for al of those to work in a geocentric universe, the model is assinine.

"You do realize that if there are any sort of observations showing Earth is definitely in motion, then relativity is an invalid theory, don't you?"

This is flatly wrong because it doesn't matter in relativity you imbecile.

Scott Reeves wrote:

"This is completely false, it has been tasted [sic] over and over and over again and it has succeeded at every point. It is tasted [sic] everyday by cellphones."

Maybe it's been tasted, as you say, but it hasn't been tested. Once you have gone a cosmologically significant distance from Earth, repeated the observations made on Earth, then and only then can you claim that relativity has been tested. Until you've done that, all that can be said at this point is that relativity has been tasted (whatever that means) and relativists have merely been gathering data from within an Earth-based reference frame, data that you all can use for comparison when you finally get in your little spaceships and zoom off to the distant stars and replicate your experiments there.

"Yeah no, it's dead by scientists because all evidence ever gathered goes against it."

Yeah, no. The only "evidence" against it is a philosophical principle. Here's another quote from The Grand Design, by Stephen Hawking:

""At first sight, all this evidence that the universe appears the same whichever direction we look in might seem to suggest there is something distinctive about our place in the universe...There is, however, an alternative explanation: the universe might look the same in every direction as seen from any other galaxy too. This, as we have seen, was Friedmann's second assumption [the cosmological principle]...But today we believe Friedmann's assumption for almost the opposite reason, a kind of modesty: we feel it would be most remarkable if the universe looked the same in every direction around us but not around other points in the universe!"

Modern scientists believe we're not at the center of the universe not based upon any empirical evidence, but upon a desire for modesty.

"Easily explainable? Look up cool hard logics [sic] video and see what explination [sic] is required for al [sic] of those to work in a geocentric universe, the model is assinine [sic]."

Between CoolHardLogic and Geocentrism, the only thing asinine is CoolHardLogic, and there's certainly no HardLogic in his obnoxious arguments against geocentrism. Guy's a jerkwad. Ad hominem attacks (that's the sort of attack I just did in the previous sentence, but I'm sure you know that, you imbecile);) against a well-meaning but misguided geocentrist who advocates an obsolete geocentric model to which no knowledgeable modern geocentrist adheres. Fernieboy100 is a straw man that CHL props up to demonstrate CHL's own misunderstanding of both modern geocentrism and relativity itself. Why don't you look up Robert Sungenis's excellent rebuttal of CHL's first video? (Yes, I know, Sungenis is a religious nut and an imbecile and you won't waste your time, right?)

"This is flatly wrong because it doesn't matter in relativity you imbecile."

Relativity itself doesn't matter because it's pseudo-science. But anyway, once again you're demonstrating your lack of understanding of relativity. If you can point to a reference frame and say, "This frame is definitely, absolutely in motion," then relativity is invalid, because you have just identified a special reference frame, unequal to other reference frames. When there is relative motion between two frames, relativity requires that each reference frame be allowed to consider itself at rest and the other reference frame in motion. If you say, "That frame is most definitely in absolute motion," then that frame cannot consider itself to be at rest and the other frame in motion. If you can point to such a frame, then you have just identified a frame that is not equal to other frames, thereby invalidating relativity. Do you really not know this, or are we just miscommunicating somewhere? You ARE familiar with the concepts of absolute motion and absolute rest, aren't you?

EmperorZelos wrote:

+Scott Reeves

Are you really so weak in your position that you have to latch onto a typo? Really? You need to both grow up then and learn some basic science.

Scott Reeves wrote:

+EmperorZelos

"Are you really so weak in your position that you have to latch onto a typo? Really?"

Not really. I only did it because I noticed elsewhere in these comments that you were very nastily cursing someone out for their atrocious spelling. Meanwhile, your responses to me have been littered with a ridiculous amount of misspelling.

"You need to both grow up then and learn some basic science."

I'm plenty grown up. If you read back through our interactions, you'll notice that you were the first to start calling me names. I ignored it at first, but lately it's getting ridiculous, so I've been returning the favor a bit. But you're right. I shouldn't be stooping to your level, so I revert to my previous position of ignoring your juvenile attacks against me personally. And I've learned plenty of basic science, and urge you to do the same. Or at least reassess whether you've actually learned what you think you've learned.

EmperorZelos wrote:

+Scott Reeves

It's only name calling if the terms are not descriptive and for you, they are very accurate. You've learned no science if you advocate geocentrism, something that is 500 years out of date.

<u>Scott Reeves wrote:</u>

+EmperorZelos

Well, then, I know what you are, but what am I infinity.

These are The Undeniable Facts (aka the CoolHardLogic-al facts), whether you choose to accept them or not:

1) A relativist can only say that Earth is in relative motion. Could be Earth that is in motion, could be the other guy. If you say something, such as the Earth, is definitely in motion, then you are in violation of relativity.

2) We, the observers on Earth, are at the center of our own observable universe. This is what the empirical evidence says. Both relativists and absolute Geocentrists acknowledge this.

3) Absolute Geocentrists say that our (we, the observers on Earth) observable universe is all that there is to the universe. There is nothing beyond our observable universe.

4) If our own observable universe is the entire universe, and we are at the center of it, the Earth can be empirically shown (interferometer experiments) to be at absolute rest. Therefore if our own observable universe is all that there is, Earth is immobile at the center of the universe.

5) According to standard Big Bang cosmology, anything beyond our (we, the observers on Earth) observable universe will be forever unobservable to us due to the expansion of space.

6) Things that are unobservable do not exist as far as science is concerned. Therefore anything beyond our observable universe does not exist as far as science is concerned. How could it, since the ability to observe something is a requirement of the scientific method?

7) If 4) is correct then relativity is false, because Earth is a preferred, special reference frame, defining an absolute rest frame.

8) The only way 4) can be false is if there is a larger universe beyond our (we, the observers on Earth) observable universe.

9) Therefore, both relativity and the Copernican principle require the existence of a universe larger than our observable universe, i.e. both depend upon something that is beyond the scope of the scientific method and rational scientific inquiry.

10) Mainstream science claims that there is a larger universe beyond our observable universe. They MUST claim this, because both relativity and the Copernican principle require it (see 9)) Google "the observable universe vs. the entire universe."

11) Relativity is presented to the public as scientific fact, yet, due to the requirement in 9), relativity is an unscientific theory.

12) Therefore, relativity is pseudo-science.

13) The only options available to us are absolute Geocentrism or relativistic geocentrism.

14) Relativity is a pseudo-science, therefore the only option available to strict adherents to science is absolute Geocentrism.

Earth is at absolute rest at the center of the universe as far as science is concerned. Relativity is a pseudo-science and anyone who adheres to it is a kook. Case closed, class dismissed.

EmperorZelos wrote:

+Scott Reeves

"A relativist can only say that Earth is in relative motion. Could be Earth that is in motion, could be the other guy. If you say something, such as the Earth, is definitely in motion, then you are in violation of relativity."

We can say everything is in motion because everything is in motion in some frame of reference. The thing is also there are certain things that shows that it's in motion such as acceleration, while the direction of acceleration and quantity differs depending on the frame of reference it can always be measured in any frame and show that it's moving and changing.

"We, the observers on Earth, are at the center of our own observable universe. This is what the empirical evidence says. Both relativists and absolute Geocentrists acknowledge this"

It's the same with every point when you look out from it, we can observe it's not stationary though.

"If our own observable universe is the entire universe, and we are at the center of it, the Earth can be empirically shown (interferometer experiments) to be at absolute rest. Therefore if our own observable universe is all that there is, Earth is immobile at the center of the universe."

Really? Cite the peer reviewed work of it. All measurements have shown that earth is not at rest. The rest is irrelevant.

Scott Reeves wrote:

+EmperorZelos

"We can say everything is in motion because everything is in motion in some frame of reference."

Incorrect. To an observer in any reference frame, that reference frame is not moving. It's the all the other reference frames that are moving. To say that you've got all the bases covered because you can always find a reference frame where any object under consideration is in motion, is to assume the viewpoint of an ultimate observer in an ultimate reference frame. Which = absolute reference frame.

"The thing is also there are certain things that shows that it's in motion such as acceleration, while the direction of acceleration and quantity differs depending on the frame of reference it can always be measured in any frame and show that it's moving and changing."

Incorrect. An observer in an allegedly accelerating reference frame can claim that it's the other reference frames that are accelerating. A car on the highway can say that he is motionless and the ground is racing past him. When he puts on his brake, a momentary gravitational field is generated ahead of the car which decelerates the entire universe. It sounds ridiculous, but that is exactly the position of general relativity. The car is always motionless from the viewpoint of its observer. It's the entire universe that accelerates, moves at a constant velocity for a time, and then decelerates.

"It's the same with every point when you look out from it, we can observe it's not stationary though."

Again, you have no proof that it's the same with every point. That's the Copernican Principle, and there's no empirical evidence that the Copernican principle is true. Again I quote Stephen Hawking from The Grand Design, pg 62: "We have no scientific evidence for or against that second assumption." The second assumption to which he refers is Alexander Friedmann's assumption, which, in Hawking's words, is "...the universe might look the same in every direction as seen from any other galaxy too." Sound even remotely like what you've been saying? As Hawking says, there's no empirical evidence for it. And Hawking is wrong that there's no evidence AGAINST it. The evidence against it is that it requires the existence of a larger universe beyond the observable universe, which is something that cannot be verified because the alleged larger universe is unobservable, and so is beyond the scope of rational scientific inquiry. So there's no evidence for the Copernican principle, and there's evidence against it. Goodbye, Copernican principle, and, through guilt by association, goodbye relativity.

"All measurements have shown that earth is not at rest."

Not ALL measurements have shown that Earth is not at rest. If ALL measurements showed that Earth is not at rest, then those measurements are showing that Earth is in absolute motion. Absolute motion is as deadly to relativity as absolute rest. All that relativity can speak to is relative motion between reference frames. An observer on Earth can say that Earth is motionless and the sun is moving. An observer on Mars can say that Mars is at rest, Earth is moving around the sun, and both are moving around the Mars. Depending on which observer you ask, Earth is either in motion, or it is stationary. That is relativity. So if you say that ALL measurements have shown that Earth is moving, you're assuming the viewpoint of a non-Earth-centered observer. And that's true. All of HIS measurements have shown that Earth is not at rest. But If you ask an observer on Earth, all his measurements will show that Earth is not moving. In relativity, you cannot say whether Earth is moving or not. You can only ask different observers about Earth's state of motion or lack of it, and all of them are correct, at least according to relativity. I don't think you are quite understanding that about relativity.

[No response from EmperorZelos to the above]

Scott Reeves later wrote (commenting on EmperorZelos's comment to another user):

+EmperorZelos

"Imma call you stupid if you keep holding onto them when I try to educate you."

Could just mean that the teacher is stupid.

EmperorZelos wrote:

+Scott Reeves

Could be, that is known to occure in life but more often than not that is not the case.

In this instance for example when the teacher says the earth is round and moving through space, the teacher is 100% absolutely correct and opposition to say otherwise is stupidity.

Scott Reeves wrote:

+EmperorZelos

You've got the round part right, but regarding the moving through space part, you're still merely a student who obviously has not yet received his diploma in Relativity.

EmperorZelos wrote:

—

+Scott Reeves

How cute! Except we know we move through space because relativity, as I bet you mean special, only applied to linear motion and not accelerating frames of references

Scott Reeves wrote:

+EmperorZelos

"How cute! Except we know we move through space because relativity, as I bet you mean special, only applied to linear motion and not accelerating frames of references"

Only an observer in a reference frame relative to which Earth is moving "knows" that Earth is moving. You're correct that special relativity doesn't apply to accelerating frames of reference. But general relativity provides no answers regarding the question of the motion of the Earth, because Earth is only an accelerating reference frame if you assume the viewpoint of an observer outside the geocentric reference frame. For a geocentric observer, the geocentric reference frame is an inertial reference frame.

For a relativist, there is no objective fact as to whether the Earth is moving. Not even in general relativity. Earth's motion and rotation, or lack of motion and rotation, depend upon which observer you ask, and every observer you ask is correct. So according to relativity, if someone in a geocentric reference frame says Earth is neither orbiting the sun nor rotating, then he/she is just as correct as an observer in a non-geocentric reference frame who says that the Earth IS orbiting the sun and rotating.

Debate Four

Scott Reeves vs. EH

EH wrote (to no one in particular):

Correct me if I'm wrong here, and I often am when I'm going off of my memory, but the MM experiment was done to demonstrate the **existence** of the aether, right?

Scott Reeves wrote:

+EH

The MM experiment was NOT done to demonstrate the existence of the aether. It was done to detect the relative motion between the Earth and the aether. Title of the 1887 Michelson/Morley paper: "On the Relative Motion of the Earth and the Luminiferous Ether."

EH wrote:

+Scott Reeves

So it merely assumed the existence of the nonexistent aether and then went on to say "how many communists were in the White House". Nice to see McCarthyism being used by morons for more than just fearmongering, I guess…

Scott Reeves wrote:

+EH

MM also assumed the existence of relative motion, and then went on to say "how many communists were in the White House."

Basically a car's speedometer is designed to detect the relative motion between the car and the road. If the speedometer fails to detect such motion, do we say that the road does not exist? Or do we say that the speedometer failed to detect any relative motion because the car is parked in the driveway?

EH wrote:

+Scott Reeves

The hell are you talking about?

Oh shit. You're a flat earther aren't you. You believe space is a liquid or something, don't you.

Scott Reeves wrote:

"You're a flat earther aren't you. You believe space is a liquid or something, don't you."

No, and no.

EH wrote:

Then why do you believe in the existence of the aether.

Scott Reeves wrote:

I don't disbelieve in the aether for the same reason people believed in it before Einstein.

EH wrote:

Because they were ignorant and made up nonsense to explain what they didn't understand? Good policy. Hey, have you ever heard about this guy called Jesus, too?

Scott Reeves wrote:

"Because they were ignorant and made up nonsense to explain what they didn't understand?"

Yes, James Clerk Maxwell, Isaac Newton, Hendrik Lorentz, Fresnel, Oliver Lodge, Tesla – these were all completely ignorant men. Most ignorant men who ever lived.

No, I don't disbelieve in the ether because I suspect that light requires a medium through which to propagate.

"Good policy. Hey, have you ever heard about this guy called Jesus, too?"

Why? Did he believe in the ether? Do you have a Bible verse you'd like to share where he talks about the ether? Or are sin and guilt weighing so heavily upon you that you'd like to hear more of the Good News? Hallelujah, brother, he is risen, amen!

Sorry, but I don't appeal to God or religion in my arguments about relativity and geocentrism, so you'll have to hope I'll amuse you in other ways.

Debate Attempt

The following is not an actual debate, merely my response to a YouTube video to which the person who created the video has not responded as of the date this book went to press. The numbers represent the exact time in the video to which I am responding.

Response to Youtube video "A Geocentrist vs. Relativity" by   **Martymer81**
https://www.youtube.com/watch?v=YN2CLeLPDps

1:22

Have you actually read the original Michelson-Morley paper? I would say no, because your 5-point outline is a complete misrepresentation of it.

According to your bullet points, the hypothesis of Michelson-Morley is (1) "Light is a wave in a medium, the aether (or ether)." The alleged consequence of this hypothesis (2) is that "The Earth moves through the aether at at least 30 km/s, the Earth's orbital velocity."

(2) is actually NOT a consequence of (1). It is an assumption independent of (1), and was explicitly labeled as an assumption in Michelson's 1881 paper, and implicitly labeled as such in Michelson-Morley's 1887 paper. Both your prediction (3) and your conclusion (5) are dependent upon the truth of that assumption.

Anyway, (1) was NOT the hypothesis of the Michelson-Morley experiment. The title of the 1887 paper was "On the Relative Motion of the Earth and the Luminferous Ether," which gives a pretty good idea of exactly what the experiment WAS about. Is the paper titled, "On the Question of the Ether's Existence" or "On the Question of Whether Light is a Wave in a Medium"? No.

The actual hypothesis of the 1887 experiment was one of Fresnel's, that "the ether is supposed to be at rest except in the interior of transparent media." It had to do with the question of whether the ether is entrained in objects moving through it, NOT with whether the ether actually existed. The existence of the ether was another assumption of the experiment.

It's even clearer in Michelson's 1881 experiment: "The result of the hypothesis of a stationary ether is thus shown to be incorrect, and the necessary conclusion follows that the hypothesis is erroneous."

The conclusion of the 1887 paper: "It appears, from all that precedes, reasonably certain that if there be any relative motion between the earth and the luminiferous ether, it must be small; quite small enough entirely to refute Fresnel's explanation of aberration." Part of Fresnel's explanation being that the ether is stationary except in the interior of transparent media.

Where exactly in the paper does it state that the conclusion of the experiment is that there is no ether? It isn't even implied. Einstein, decades later, was the one who said there was no ether.

The point: your 5-point outline of MM is riddled with errors and a conclusion based upon a biased misstatement of the true experimental hypothesis.

Here is a summary of the experiment inferred from your 5-point outline of it:

Light is a wave in a medium, the aether (or ether). Therefore the Earth moves through the aether at at least 30 km/s, the Earth's orbital velocity. The experiment failed to detect Earth's motion through the aether. Therefore the aether does not exist.

Completely illogical. Neither of the "therefores" follows from the assertion in their respective preceding sentences.

A more accurate summary of the experiment, NOT a summary of the experiment as re-interpreted in hindsight by relativists to support their theory (which is what your 5-point summary is), is:

The aether is supposed to be at rest except in the interior of transparent media (according to Fresnel's theory). Assuming the Earth moves against the aether at at least 30 km/s, the experiment will detect such motion. The experiment failed to detect such motion. Therefore, Fresnel's explanation of aberration (stationary ether -- NOTE: NOT NO ether) is entirely refuted.

You have misrepresented/misstated/mischaracterized the entire experiment to fit your biased interpretation of the results. Use it as evidence in favor of relativity if you'd like, but at least give an honest summary of the experiment. Is your position so weak that it can't withstand an honest summary?

1:36

"And you do realize that the experiment has been repeated with more sensitive equipment, right?"

So they just have more accurate data that there is no relative motion between the Earth and the ether. Or that Fresnel's hypothesis of a STATIONARY ether is incorrect. So what? Repetition of the same experiment with more sensitive equipment does not change the hypothesis of the original experiment or its conclusion. An increase in the accuracy does not change the fact that the experiment is only designed to detect the relative motion of the Earth and the ether, NOT to detect whether the ether actually exists. You DO realize this, right? As I said to some other commenter, the failure of a car's speedometer to detect relative motion between the car and the road is not evidence that there is no road. It is evidence that there is no relative motion. Anything more is your own explanation of WHY there is no relative motion.

But let's pretend you're right. The ether has been disproven. Why then do they keep repeating the experiment? Educational purposes? To make sure there's still no ether 130 years later? You've got to watch out for that ether stuff, because some of it might have condensed out of the void after the original experiment? Whatever the reason, assume for the sake of argument that it's no longer about the ether. Even so, as you say, they're still testing for the effect of motion on the measured speed of light, still testing for evidence of c −v, v being the velocity of Earth through space. And have they detected any evidence of c-v? No. Hypothesis of a motionless Earth: supported, regardless of the existence or non-existence of the ether.

2:05

"Uh, no, it [relativity] has its limits, indicating that it's incomplete, but it works perfectly fine within those limits..."

Perfectly fine? No. Dark matter is evidence that it doesn't work perfectly fine.

Anyway, perhaps when it's complete and we have a ToE, we'll see that the reason relativity was incomplete was because it assumed there are no preferred reference frames. Except then we'll see that it wasn't actually incomplete. It was wrong.

2:33

"Nope. The Twin Paradox is a misunderstanding within special relativity."

Nope, it's not. It's a thought experiment within special relativity. The thought experiment itself is not a misunderstanding; rather, people who think it can't be resolved have a misunderstanding of the supposed paradox.

Bowden calling it a paradox isn't an indication of his

misunderstanding; it's an indication that he's choosing to call it by the common name for that thought experiment. So simply referring to it as "The Twins Paradox" doesn't mean someone doesn't understand that it isn't actually a paradox, any more than calling a cheese product "cheese" doesn't mean it isn't understood that it isn't really cheese; it simply means that you're referring to it by one of its accepted names. You have to look for other evidence that Bowden doesn't understand The Twins Paradox.

7:03

"And what would that material be, and what other effects would it have that we could look for and confirm its existence?"

I find it laughable that you're dismissing Poor's idea of a small amount of material around the sun simply because we don't know what the material would be and need some other effects to confirm its existence, yet you're willing to accept the existence of an invisible, unknown, mysterious and purely hypothetical type of matter to explain the galaxy rotation problem. Double standard. Seems like there might be some dark matter around the sun. Dark matter. Wonderful stuff. We can use it to explain any discrepancy between theory and observation.

8:19

"When you're standing in a field and looking up at the stars... The point of general relativity is that there is no correct way of looking at it. Are you spinning or are the stars orbiting you?"

Thank you for acknowledging the equality of a geocentric reference frame, be it absolute or relative. Getting some relativists to do so is a monumental task.

9:36

"You do know that relativity is a classical theory, right? Classical physics includes everything except quantum mechanics."

Only according to people who hold to a such a definition of classical physics. Many people (not just us alleged crackpots) consider classical physics to be physics pre-1900, excluding relativity.

For example:

http://dictionary.reference.com/browse/classical-physics
http://www.merriam-webster.com/dictionary/classical
https://www.lhup.edu/~dsimanek/ideas/allabout.htm
http://www.thefreedictionary.com/classical+physics

10:26

"I've looked for a peer-reviewed paper by Setterfield and Barnes and I can't even find a reference to it."

You DO know how to competently do an Internet search, don't you? It's a valid question, because I didn't know who Bowden was referring to either, but half a minute's work on Google reveals that the two people referred to are creationists, who separately have written several articles, some of which Bowden has apparently read, since he cites them.

I doubt that your own apparently feeble research skills revealed as much, because if they had, I'm sure you wouldn't have wasted the opportunity to mock the fact that Bowden is citing non-peer-reviewed, creationist sources (no disrespect to creationists intended). Simply Google "Setterfield and Barnes muons" and boom! You have plenty of stuff to work with. One of the guys cited even has his own entry on Wikipedia, so it's not difficult to find out who they are. How did your search fail to uncover this? They're clearly creationists. You have done a disservice to your audience by not revealing this and mocking Bowden for citing creationists. I DOUBT you failed to mention this merely to grant Bowden a sole merciful moment of reprieve from your scorn, so the only conclusion is that your research skills are ineffective, OR that you didn't actually look, despite your claim that you did.

"If you're going to cite sources, why don't you provide proper references so that I can check for myself."

I'll second that, because you apparently can't dig information up on your own without having someone take you by the hand and lead you directly to it. Bowden gave you two names and a few other keywords to work with, and you came up with NOTHING?! I came up with plenty of stuff in about thirty seconds. No peer-reviewed papers, of course, but geez, I can't actually believe you wouldn't have jumped at the chance to expose Bowden's use of creationist sources.

Or maybe you only searched the websites of a few peer-reviewed journals, deigning to forsake the common Internet search. Come on! You should have known better.

11:09

"Also, every time we use a GPS device, we use technology that wouldn't work if the speed of light were different in different directions."

Incorrect. GPS is based upon the average speed of light on a round trip, which doesn't preclude the possibility of an anisotropic speed of light on each leg of the journey. The one-way velocity of light has never been measured, so it's only by convention that the speed of light is not different in different directions, not by any empirical evidence. What we know as the speed of light is the average speed over a two-way path. Also, absolute Geocentrism predicts exactly the same necessary corrections to the GPS system that relativity does.

So let's go set up a GPS system on some other planet and see if it works identically to the way it does on Earth. Earth: the only place in the universe where GPS will work as it does. Lo and behold, where is the only place in the universe where a GPS system has been set up? You guessed it: Earth.

GPS cannot yet be validly claimed as conclusive support for relativity that excludes all contending theories, since it supports absolute Geocentrism equally well.

11:57

"And these [Sagnac's fringe shifts] are always the same regardless of how fast you're moving or in what direction."

Not true. The magnitude of the fringe shifts depends upon the relative rotational speed of the table. The faster the table or the universe around it rotates, the greater the fringe shift.

12:14

"You're citing an example of how the non-existence of the ether is used for practical purposes."

No, he's citing an example of how the unequal lengths of the two light paths around the table (unequal due to rotation from the viewpoint of an observer external to the frame of the device) are used for practical purposes by an observer.

Assume you are correct and there is no ether. In the Sagnac experiment, from the viewpoint of an observer on the table, the table is not rotating and he is stationary relative to his interferometer, and the light paths are equal, yet he still measures a fringe shift. However, in the Michelson-Morley experiment, the observer is likewise stationary relative to the device and the light paths are likewise equal, yet he measures NO fringe shift (except that he does, but we'll put it down to the experiment's margin of error). How do YOU explain this inconsistent result?

12:37

"How people reacted to the theory before the overwhelming evidence that we have today was available has no impact on its validity."

You're right. It doesn't. But you also don't know how they would have reacted if they DID have all the "overwhelming" evidence. You seem to be implying that they would have reacted differently, but you can't know what might have been.

As for the "overwhelming" evidence gathered by observers from within a geocentric reference frame -- Michelson-Morley, GPS, particle accelerators, et al. Go out into the rest of the universe, set up these things, and replicate the results you obtained on Earth. THEN you can claim them as support for relativity. For now, they only support a geocentric reference frame.

12:55

"Right. Because you can't ignore one of the cornerstones of physics and still be taken seriously as a physicist."

Right. Just as a self-proclaimed Satan-worshipper wouldn't be taken seriously as a Catholic priest. You've got to drink the Kool-Aid of whichever group you want to join. Which was Bowden's point. But just because someone with contrary views is barred from a certain group doesn't mean that the views of that group are correct.

13:30

Except that it's not [false]. It's perfectly valid at macroscopic scales, and that's where it's applied."

Which is exactly what you would expect someone who advocates a pseudoscience to say. Relativity is NOT "perfectly valid at macroscopic scales." If it were, you wouldn't need to fabricate dark matter and dark energy in order to reconcile theoretical prediction with empirical observation, two substances for which the only "evidence" is exactly the prediction/observation problem you're attempting to solve.

14:21

"The Earth is only the center of the observable universe, and that's only for the trivial reason that we must observe the universe from where we happen to be. Far from land, your boat is always the center of the observable ocean."

You're implying that there is a larger universe beyond the observable universe. What empirical evidence do you have (1) that this hypothetical larger, unobservable universe actually exists, and (2) that we are not at the actual center of that larger universe? Prove your implicit claim that our observable universe is bounded by a horizon rather than an edge. Also, what physical law says that the Earth, or any other planet in the universe for that matter, MUST be the metaphorical equivalent of a boat FAR FROM LAND at the center of its own observable ocean, utterly precluding the possibility of any boat being closer to one shore or another and observably NOT at the center of its observable ocean?

[insert Martymer81-style video clip of me making condescending faces at the camera, playing with my iPhone as I patiently wait with crickets chirping in the background]

16:12

"...so you have made the decision not to care about reality."

Says Martymer81 as he studies a CGI animation of gravitational lensing. That's classic.

Debate Five

Scott Reeves vs. NM

Dmj42 dexron wrote:

Einstein? Heliocentric theory was around long before Einstein, and for good reasons. Geocentrism is ridiculously complicated AND doesn't match observation (and doesn't make a lot of sense, either, unless you posit that God is intervening every second to keep everything in place).

Scott Reeves wrote:

"Geocentrism is ridiculously complicated AND doesn't match observation"

It's not ridiculously complicated. What's ridiculously complicated about it? And actually, absolute Geocentrism matches ALL observations. There's a reason why modern scientists say that the Earth is at the center of the observable universe: because it is. But unfortunately for those same scientists, there is absolutely no evidence of anything beyond our observable universe. So to say that Earth isn't at the center of the actual, entire universe is to go against all the observational evidence. Which observation do you say that Geocentrism doesn't match?

"(and doesn't make a lot of sense, either, unless you posit that God is intervening every second to keep everything in place)."

Geocentrism makes as much sense as heliocentrism, and you DON'T have to posit that God is intervening every second to keep everything in place in a geocentric model any more than you have to posit the same thing regarding a heliocentric model, or a galacto-centric model, or whichever -centric model you want to choose. Do relativistic geocentrists have to claim that God is intervening every second to keep everything in place? No? Then why do absolute Geocentrists?

NM wrote:

"There's a reason why modern scientists say that the Earth is at the center of the observable universe"

HAHAHAHAHAHAHAHAHAHAHAHAHAHA

Scott Reeves wrote:

"HAHAHAHAHAHAHAHAHAHAHAHAHAHA"

Why are you laughing maniacally? Have you never heard modern scientists say that every point in the universe will see itself as the center? This is an acknowledgement that Earth looks like it's at the center of the universe to an observer on Earth. But scientists have no evidence to back up their claim that EVERY point will see itself as the center. Have we BEEN to any other point in the universe and made such observations? Thus far, the available evidence only supports the view that we are at the center of the entire universe. It can't do anything but, since all the evidence has been gathered by observers on Earth. Anything more is nothing but an assumption and a desire to not be in a special place in the universe. The reason you're laughing is because you have absolutely no means other than ridicule to refute me on this point.

NM wrote:

There is a difference between saying "everywhere is the center of the universe" and saying the entire universe spins around the earth. They simply mean the universe has no center. All credible physicists accept relativity as truth, and relativity disagrees with the geocentric model, heavier bodies don't orbit lighter bodies.

Scott Reeves wrote:

"There is a difference between saying 'everywhere is the center of the universe' and saying the entire universe spins around the earth."

Yes there is. And the difference is that there is observational evidence for the latter but not the former.

"They simply mean the universe has no center."

That may be what they mean, but the fact is that they acknowledge that Earth is at the center of its own observable universe. They acknowledge this, because all observational evidence shows that we are. They then hypothesize that EVERY point in the universe is the center of its own observable universe. An inherent part of this hypothesis is that there is a larger universe beyond Earth's observable universe. Scientists have yet to test this hypothesis. Until such testing has been done, all observational evidence shows that Earth is at the center of the universe, period.

"All credible physicists accept relativity as truth,"

Which only means that they aren't all that credible.

"and relativity disagrees with the geocentric model,"

Actually, Relativity does NOT disagree with the geocentric model. It merely disagrees with the ABSOLUTE Geocentric model. Relativity MUST have its own geocentric model of the universe, else Relativity is invalid. Einstein knew this, which is why he spent time defending the geocentric reference frame.

"heavier bodies don't orbit lighter bodies."

Heavier bodies don't orbit lighter bodies in a geocentric universe either. The sun is technically not orbiting the Earth. It is orbiting the barycenter of the entire universe, in the same way that in the heliocentric model, the planets orbit the barycenter of the solar system rather than the sun itself. Even the sun orbits the barycenter of the solar system in a non-geocentric model. No credible geocentrists claim that Earth is massive enough to gravitationally hold the entire universe in orbit around it.

Debate Six

Comments from the YouTube video "TYCHO BRAHE Says No Spheres NoParallax No Planets - All Lies" by jeranism

Scott Reeves wrote (responding to MomoTheBellyDancer's comments to Last Trump):

"That is not an assumption [that the Earth is revolving around the sun]. The fact that we can observe stellar aberration is already plenty of evidence. Another piece of evidence is the fact that Newtonian physics perfectly describe the motion of planets, including earth."

We can also observe and explain stellar aberration from a geocentric frame. Stellar aberration doesn't speak to whether Earth is revolving around the sun.

As for Newtonian physics, choosing to make Newtonian calculations using a non-geocentric coordinate system does not mean that the Earth is actually revolving around the sun. It simply means that you've chosen to make Newtonian calculations using a coordinate system in which the Earth is revolving around the sun.

As for Newtonian physics perfectly describing said motion of the planets – they didn't perfectly describe the precession of Mercury, did they? So thy DON'T "perfectly" describe the motion of the planets.

"No, because you have to introduce massive amounts of unknown variables to make the geocentric model work, which simply disappear then you go with the heliocentric model. Occam's razor compels us to go with the model with the least amount of assumptions."

Occam's razor is a philosophical preference for simplicity, not a physical law that governs the universe. The geocentric frame IS just as valid as any other frame, unless you want to deny Relativity. What you're basically saying is that Occam's razor compels us to conclude that all reference frames are not physically equivalent, in violation of Relativity.

Perhaps you might object that you were referring to the geocentric MODEL as being invalid, not the geocentric REFERENCE FRAME. But how can you acknowledge the geocentric reference frame yet deny the geocentric model that goes with it? If you're going to allow someone to assume the role of an observer within the geocentric reference frame, then that observer MUST have a model that describes the universe from his geocentric viewpoint, and that model MUST be as valid as any other model. If that model is not fully developed by such an observer, it MUST be possible to fully develop it, or else Relativity is an invalid theory. And I'm assuming you are not an anti-relativist.

MomoTheBellyDancer wrote:

"Stellar aberration doesn't speak to whether Earth is revolving around the sun."

Then explain how we could get stellar aberration that way.

"choosing to make Newtonian calculations using a non-geocentric coordinate system does not mean that the Earth is actually revolving around the sun."

Then present a model in which the Newtonian calculations are correct, but in which the earth does *not* orbit the sun. Make sure it is at *least* as comprehensive as the heliocentric model, with as few assumptions as possible.

"Occam's razor is a philosophical preference for simplicity, not a physical law that governs the universe."

Irrelevant. It states that we should choose the model that works fine with the least amount of assumptions. Really, why would we throw in a lot of assumptions when a simpler model explains the same facts just as well, if not *better*?

"The geocentric frame IS just as valid as any other frame, unless you want to deny Relativity."

In the geocentric model, the universe would be revolving the earth at several trillion times the speed of light.

"What you're basically saying is that Occam's razor compels us to conclude that all reference frames are not physically equivalent, in violation of Relativity. "

No, Occam's Razor compels us to conclude that they *are* equivalent. You're not using it correctly.

"Disclaimer: these comments should not be construed as support for Flat Earth or a denial of stellar parallax or of accepted standards for distances to the sun and stars."

You could've fooled me.

Scott Reeves wrote:

"Then explain how we could get stellar aberration that way."

The fact is that if you are a relativist and you believe that stellar aberration cannot be explained from within a geocentric reference frame, then you are actually an anti-relativist, because you do not believe that all reference frames are physically equivalent. If you ARE a relativist, then you MUST admit that stellar aberration can be explained from within a geocentric frame, even if you don't currently know what the explanation is. So I don't need to explain it, because neither true relativists nor absolute Geocentrists (this one, at least) disagree that stellar aberration is geocentrically explicable.

That being said, here is one of several explanations put forward by modern Geocentrists: "Stellar aberration is star motion centered on the sun as viewed from Earth, hence there is no aberration in stellar motion as seen from the sun. The aberration is due to the apparent shift in the stellar positions that are centered on the sun. This is a parallax effect due to the change in position of a reference point." - Robert Sungenis, Galileo Was Wrong, the Church Was Right, Volume 1.

And anyway, to quote Stephen Hawking in The Grand Design: "So which is real, the Ptolemaic or the Copernican system? Although it is not uncommon for people to say that Copernicus proved Ptolemy wrong, that is not true. As in the case of our normal view versus that of the goldfish, one can use either picture as a model of the universe, for our observations of the heavens can be explained by assuming either the earth or the sun to be at rest."

Stellar aberration qualifies as one of the "observations of the heavens," and can thus be explained by geocentrism, according to the man who is allegedly one of the greatest physicists of our time. Unless, of course, he meant to say all our observations except the aberration of light. Which could be possible, since he is incorrect that our observations can be explained by the Ptolemaic model. Modern geocentrists do not adhere to the Ptolemaic model, because we know it is inaccurate, as it has all the planets orbiting the Earth. But I'm sure Hawking meant a geocentric model, rather than the Ptolemaic model specifically.

"Then present a model in which the Newtonian calculations are correct, but in which the earth does not orbit the sun. Make sure it is at least as comprehensive as the heliocentric model, with as few assumptions as possible."

I present to you: the geocentric model! Ta-da! Could be a relativistic geocentric model, or it could be an absolute Geocentric model. I'm not going to explain the details of the model because there's probably not enough space in this comments section, and because frankly I as an absolute Geocentrist do not KNOW all the nuances of the absolute Geocentric model any more than relativists appear to know all the nuances of their own relativistic model As far as Newtonian calculations – as I pointed out, Newtonian calculations aren't even correct in the model in which Earth DOES orbit the sun (precession of Mercury, for example). But I assume you mean that Newton's laws don't allow for a greater mass like the sun to orbit a lesser mass like the Earth, and so the geocentric model doesn't work with Newtonian calculations because of that. But the sun only appears to be orbiting the Earth in a geocentric reference frame; the sun is actually orbiting the center of mass of the universe, and the Earth happens to be located at that center. A system of masses orbits the center of mass of the system, according to Newton, and that is not violated in a geocentric model.

"Irrelevant. It states tat we should choose the model that works fine with the least amount of assumptions. Really, why would we throw in a lot of assumptions when a simpler model explains the same facts just as well, if not better?"

To what assumptions are you referring? And as far as choosing a simpler model, simpler does not necessarily mean correct. If your allegedly simpler model is too simple to accommodate a complex universe, then the simpler model is not preferable because it doesn't represent reality.

"In the geocentric model, the universe would be revolving the earth at several trillion times the speed of light."

This argument is often presented against geocentrism, but it is not valid. Einstein, Dialogue About Objections Against the Theory of Relativity: "The situation, that the fixed stars are circling with tremendous velocities, when one bases an examination on such a coordinate system, does not constitute an argument against the admissibility, but merely against the efficiency of this choice of coordinates..." See, Einstein was actually a geocentrist, as all relativists are. We're ALL geocentrists; the only difference between us is whether we claim that all reference frames are physically equivalent.

Phil Plait, Seriously? Geocentrism?: "Things actually can move faster than light relative to the coordinate system, it's just that things cannot move past each other with a relative speed greater than light. In the weird geocentric frame where the Universe revolves around the Earth, that is self-consistent. In other words, the Neptune-moving-too-quickly argument sounds good, but in reality it doesn't work, and we shouldn't use it."

I'm a quote miner. Should I continue throwing out similar quotes, or are those two sources reputable enough?

"No, Occam's Razor compels us to conclude that they are equivalent.

You're not using it correctly."

So now Occam's Razor, as well as Relativity, compels us to conclude that the geocentric frame IS equivalent to the heliocentric?

"You could've fooled me."

Only if you believe that Flat Earth and geocentrism are not completely separate and unconnected theories (and I can understand why you might have that impression, since a lot of flat Earthers seem to be trying desperately to connect the two), and that stellar parallax and accepted standards for distances to the sun and stars can't be explained from within a geocentric reference frame.

MomoTheBellyDancer wrote:

"Robert Sungenis, Galileo Was Wrong, the Church Was Right, Volume 1."

Not a peer-reviewed work. Dismissed.

"Stephen Hawking in The Grand Design"

Not a peer-reviewed work. Dismissed.

"the sun is actually orbiting the center of mass of the universe, and the Earth happens to be located at that center."

With would mean that the earth would be at the bottom of an

immense gravity well i.e. a *black hole*. This is not what we observe. Dismissed.

"To what assumptions are you referring?"

What would cause the retrograde motions of planets in the geocentric model? Oh right, you will have to make the planets go loop-de-loop around some unexplained center.

OR you could put the sun be the center of the solar system.

Geocentric model -> dismissed.

"simpler does not necessarily mean correct."

Whether model is correct is irrelevant. What *is* relevant is whether it explains all observations with the least amount of assumptions.

"If your allegedly simpler model is too simple to accommodate a complex universe"

Then it wouldn't explain all observations, so your objection is silly.

"The situation, that the fixed stars are circling with tremendous velocities, when one bases an examination on such a coordinate system, does not constitute an argument against the admissibility, but merely against the efficiency of this choice of coordinates"

And later he says, "For the decision which representation to choose only reasons of efficiency are decisive, not arguments of a principle kind."

This is still a strike against geocentrism. Besides, it's not a peer-reviewed article, so it's dismissed anyway

"Einstein was actually a geocentrist"

Don't be stupid.

"Phil Plait, Seriously? Geocentrism?"

Not a peer-reviewed work. Dismissed.

"I'm a quote miner."

That's not something to be proud of.

"So now Occam's Razor, as well as Relativity, compels us to conclude that the geocentric frame IS equivalent to the heliocentric?"

Point me to a peer-reviewed work that posits this, and then we'll talk.

Scott Reeves wrote:

"With would mean that the earth would be at the bottom of an immense gravity well i.e. a black hole. This is not what we observe.

Dismissed."

Really? Google 'Are we inside a black hole' and see how many articles come up about how scientists on your side are proposing that we are inside a black hole.

"What would cause the retrograde motions of planets in the geocentric model? Oh right, you will have to make the planets go loop-de-loop around some unexplained center."

What causes it? The planets orbiting the sun, even as the universe as a whole revolves around the universal barycenter. There's nothing inexplicable about retrograde motion. It's a combination of gravity and the rotation of the universe. What causes the retrograde motion in the relativistic geocentric model?

"OR you could put the sun be the center of the solar system."

You could, but for Relativity to be a valid theory, you also have to be able to put the Earth at the center. If you're going to advocate for Relativity, then you can't deny the geocentric reference frame.

"Whether model is correct is irrelevant. What is relevant is whether it explains all observations with the least amount of assumptions."

How does the heliocentric model explain the fact that neither Newton nor Relativity can explain why galaxies are spinning faster than they should according to either theory, yet the galaxies don't fly apart? Neither theory explains it. They assume the existence of dark matter and dark energy, neither of which has been observed to date. So none of the models you are advocating explain all observations.

"Then it wouldn't explain all observations, so your objection is silly."

And as I noted above, your allegedly simpler model does not explain all observations so your objection to my objection is equally silly. Hypothesizing the existence of an unobserved entity to explain disagreements between your theory and observation is actually no explanation at all.

"This is still a strike against geocentrism. Besides, it's not a peer-reviewed article, so it's dismissed anyway."

It's only a strike against geocentrism if you believe that Occam's Razor is infallible and utterly applicable to every situation imaginable. Einstein does not say that it's impossible for the universe to rotate around the Earth at several trillion times the speed of light, which is what your previous statement implied.

Dismissing highly reputable scientists who disagree with your statement regarding geocentrism being forbidden due to some faster-than-light problem simply because they're not peer-reviewed statements does not change the fact that highly reputable scientists are contradicting you.

Plus, one of those contradictory voices is Einstein himself, and he's got to be the most peer-reviewed person I could possibly quote.

Everything he has written has been intensely scrutinized for a hundred years. Point me to a peer-reviewed article that says the statement I quoted is not accurate, or that the geocentric reference frame is not valid within relativity.

"Don't be stupid."

I'm not being stupid. Are you saying that Einstein was NOT a geocentrist? A geocentrist from a relativistic perspective is simply someone who assumes the viewpoint of an observer within an Earth-centered frame. All relativists have to allow for the possibility of assuming such a viewpoint. Therefore, all relativists are inherently geocentrists. The difference between relativistic geocentrists and absolute Geocentrists is that relativistic geocentrists claim that the geocentric reference frame is simply one of a multitude of equal of physically equivalent reference frames. If you claim to be a relativist and deny the geocentric reference frame as being equally valid, then you are not actually a relativist.

"That's not something to be proud of."

And yet I am.

"Point me to a peer-reviewed work that posits this, and then we'll talk."

I wasn't positing this. In your earlier comments, you first said that Occam's Razor compelled us to reject geocentrism, and then later said that Occam's Razor compelled us to conclude that all reference frames were physically equivalent. I was merely questioning your apparent inconsistency.

MomoTheBellyDancer wrote:

"Really?"

Really.

"Google 'Are we inside a black hole'"

I knew you'd mention this. But of course, it's nonsense (as most anything else you say). Yes, the *whole universe* could be inside of a black hole (or equivalent), but that's not the same as the earth being in a gravitational well with the rest of the universe rotating around it.

"What causes the retrograde motion in the relativistic geocentric model?"

The fact that all planets go around the sun in elliptic orbits. Geocentric model -> still dismissed.

"If you're going to advocate for Relativity, then you can't deny the geocentric reference frame."

Nonsense.

"How does the heliocentric model explain the fact that neither

Newton nor Relativity can explain why galaxies are spinning faster than they should according to either theory, yet the galaxies don't fly apart?"

Citation needed.

"your allegedly simpler model does not explain all observations"

Citation needed.

"It's only a strike against geocentrism if you believe that Occam's Razor is infallible and utterly applicable to every situation imaginable."

Please explain why, given the choice between two models that explain observations equally well, we should *ever* have a reason to choose the model with the most assumptions.

"Einstein does not say that it's impossible for the universe to rotate around the Earth at several trillion times the speed of light"

No, he doesn't. Einstein merely points out that this motion in and of itself it not impossible *in principle*, but that it is still impossible *in practice*, since nothing can move at trillions the times the speeds of light. Try to read for comprehension. Better yet, try to read the original German.

"Dismissing highly reputable scientists who disagree with your statement"

Einstein doesn't disagree with my statement. You're quoting him out of context.

"Plus, one of those contradictory voices is Einstein himself, and he's got to be the most peer-reviewed person I could possibly quote."

His *articles* are peer-reviewed, not his person.

"Are you saying that Einstein was NOT a geocentrist?"

If you claim he was, then I will need a citation. Something along the lines of Einstein saying "I believe the earth is unmovable in the center of the universe" will do.

"And yet I am."

And that's why you're stupid.

"I was merely questioning your apparent inconsistency."

There is no inconsistency.

Scott Reeves wrote:

"I knew you'd mention this."

You, sir, are a true seer.

"But of course, it's nonsense (as most anything else you say). Yes, the whole universe could be inside of a black hole (or equivalent), but that's not the same as the earth being in a gravitational well with the rest of the universe rotating around it."

Fine. The geocentric reference frame is invalid because otherwise we would all be crushed to death in the singularity of a black hole. And yet we're alive in a geocentric reference frame. Go figure. So I guess something must be wrong with the theory of black holes as currently formulated. Or something wrong with your assertion that Earth being in a gravitational well with the rest of the universe rotating around it would mean that we are inside a black hole. Because it certainly can't mean that a geocentric reference frame is invalid, because that would mean relativity is invalid.

"The fact that all planets go around the sun in elliptic orbits."

Which is what they do in a geocentric model as well. All planets except Earth, that is. Unless you're going with some old school Ptolemaic model to which the majority of modern absolute Geocentrists no longer adhere. The way the planets are observed to move against the celestial sphere does not mean that geocentrists must claim the planets do not orbit the sun in elliptic orbits, any more than it means such a thing for heliocentrists.

Geocentric model not dismissed.

"Nonsense."

Then you don't quite understand relativity.

"Citation needed."

Citation needed for what? The question where I asked for your explanation, my assertion within the question that galaxies are spinning faster than they should according to either theory, or my assertion that neither relativity nor Newtonian dynamics can explain that faster spin?

If you're asking for citation on the question itself, I don't need to cite anything to ask you a question. If you're asking for citation about the question's assertion that galaxies are rotating faster than they should be according to current theories, it's standard knowledge and really shouldn't need any citation. But if you want citations, go to the wikipedia entry on 'galaxy rotation curve' and browse through the articles in the footnotes. I'm confident that one of them supports me. As for whether Relativity or Newton's laws can explain this rotation problem, the hypothesis put forth as an explanation is dark matter. But a hypothesis that has not been tested is not a valid explanation. And no, saying that the galaxy rotation problem is both the test for and evidence of dark matter is not an explanation either. It is circular reasoning.

The fact that the concept of dark matter exists and is widely accepted by the mainstream scientific community is all the citation I need to back up what I said in my earlier comment. And if you dispute that the concept of dark matter exists and is widely accepted by the mainstream scientific community, then all I can say is, "Have you been living under a rock for the past few decades?"

"Citation needed" [to my comment that 'your allegedly simpler model does not explain all observations']

Dark matter.

"Please explain why, given the choice between two models that explain observations equally well, we should ever have a reason to choose the model with the most assumptions."

Possibly we wouldn't ever have such a reason, but I can't really address this unless you give me a list of the assumptions of both models which show that one model has more assumptions than the other. And the list for each has to contain only assumptions that are not special pleadings favoring one model over the other, or that are not themselves assumptions that the opposing model is correct.

"No, he doesn't. Einstein merely points out that this motion in and of itself it not impossible in principle, but that it is still impossible in practice, since nothing can move at trillions the times the speeds of light. Try to read for comprehension. Better yet, try to read the original German."

Yes, he does. And he doesn't say it's impossible in practice. He says it's inefficient in practice to use a coordinate system based upon such a reference frame, but that the tremendous speed of the circling stars is not a problem when you examine the geocentric coordinate system. And he's right. Unless you're some huge fan of mathematics who enjoys a challenge, then using a simpler coordinate system would be more efficient because it wouldn't take as much time and you'd have less chance for error creeping in. But mathematical inefficiency does not render a reference frame invalid. As for the original German, does the German concept of 'impossible' translate to the English word 'inefficient'? I don't know, but I'm going to assume no. Besides, Einstein is not the only one who talks about the speed of light not being a problem for a geocentric frame. I mentioned Phil Plait for one, but there are plenty of others. The whole faster-than-c problem that's bandied about by CoolHardLogic and others is actually no problem at all. It's only a demonstration that those people have not actually dug deeply enough into the geocentric issue, and don't know what they're talking about.

"Einstein doesn't disagree with my statement. You're quoting him out of context."

Actually he does, and no I'm not. He is actually defending the validity of the geocentric reference frame, because he understands that if the geocentric reference frame can be shown to be invalid in any way, then so is relativity, since ALL reference frames must be equally valid for his theory to be valid.

"His articles are peer-reviewed, not his person."

Yes, but his peer-reviewed articles are so thoroughly peer-reviewed, verified and accepted by the scientific community that it's pretty safe to assume that something he says regarding relativity outside a peer-reviewed article is most likely correct. Anyway, I do not subscribe to relativity, so I really don't care whether Einstein is correct or not. As an anti-relativist, I hold that he most certainly is

NOT correct. My only reason for mentioning him and other relativists is to get anti-geocentrists such as yourself to realize that relativity MUST accept the validity of a geocentric reference frame. People who understand relativity realize that the geocentric frame is valid, which is why you get people like Einstein and Hawking defending it. It's merely one equal reference frame among a multitude of reference frames, according to them. If you can't accept that, then we haven't even actually gotten to the point yet of debating whether the geocentric reference frame is relative or absolute. Not to mention that if you can't accept any sort of geocentric reference frame, then you are both an anti-relativist and an anti-absolutist, and since relativity and 'absolutivity' are the only possibilities, rejection of both means that you are going entirely against reason and logic.

"If you claim he was, then I will need a citation. Something along the lines of Einstein saying 'I believe the earth is unmovable in the center of the universe' will do."

To the best of my knowledge, he never made such a direct and comprehensive statement, and even if he had, he would not have made it unconditionally. If he made such a statement unconditionally, he would be claiming to be an absolute Geocentrist, which I acknowledge he never did. He only espoused a democracy of reference frames and defended a geocentric frame specifically, making him implicitly and logically a geocentrist. But only relatively. He was also a Mars-centrist, a heliocentrist, a Moon-centrist, a Tatooine-centrist, a point-in-deep-space-centrist, etc. Whatever sort of observer the situation called for, he was it. For a relativist, any point in the universe can be chosen as the center of a relatively immobile reference frame. It is modern scientists who add fuel to relativity's funeral pyre by claiming that we are at the center of our observable universe.

"And that's why you're stupid."

That's fine. I don't think you're stupid. But also, thanks for taking the time to respond to my comments. I appreciate the opportunity to be challenged and forced to defend my position. Seriously.

"There is no inconsistency."

There was, even if it was unintentional. But I'll just put it down to miscommunication on your part, and you can put it down to misunderstanding or downright stupidity on my part.

MomoTheBellyDancer wrote:

"And yet we're alive in a geocentric reference frame."

I am not saying that you can't have a geocentric *reference frame*. I am saying that the earth is not at the center of the universe with he univere orbiting around it. For that, the vast majority of the universe would have to break the speed of light limit.

We'd also be dead.

"Which is what they do in a geocentric model as well. All planets except Earth, that is."

If all planets except for the earth wold orbiting the sun, then we'd seem them move in completely different patterns across the sky. The patterns we *do* see though match with *all* planets orbiting the sun, *including the earth.*

"Citation needed for what?"

Your assertion that "neither Newton nor Relativity can explain why galaxies are spinning faster than they should according to either theory."

"it's standard knowledge"

No, it's not.

"really shouldn't need any citation"

It does, so get to it.

"the hypothesis put forth as an explanation is dark matter."

Dark matter has nothing to do with the claim that the "galaxies should fall part due to their rotation." Really, where do you get this stupid idea?

"Dark matter"

Dark Matter falls perfectly within Newtonian mechanics. It's because of that fact that we can detect its presence, since it has a definitive gravitational influence on surrounding matter.

"Possibly we wouldn't ever have such a reason"

Good.

"And he's right."

No, you are *making* him right by misrepresenting his position. As long as you keep doing that, I will ignore the rest of your blather, since it's nothing but a dishonest move anyway.

"To the best of my knowledge, he never made such a direct and comprehensive statement"

There you go.

"That's fine. I don't think you're stupid."

Well, *I* am not the one doing the quote-mining.

"There was, even if it was unintentional."

You failed to point out any inconsistency. It's *you* who has no idea what a reference frame really is.

Scott Reeves wrote:

"I am not saying that you can't have a geocentric reference frame."

At last, forward movement.

"I am saying that the earth is not at the center of the universe with he univere orbiting around it."

Fair enough. But you do realize that the majority of modern scientists claim that Earth is at the center of our observable universe?

"For that, the vast majority of the universe would have to break the speed of light limit."

No it wouldn't, not according to Einstein and many others. We've already been over this. But if you fail to accept that, we can agree to disagree.

"We'd also be dead."

And yet we're not.

"If all planets except for the earth wold orbiting the sun, then we'd seem them move in completely different patterns across the sky."

Citation needed.

"The patterns we do see though match with all planets orbiting the sun, including the earth."

Correct. If you assume the role of an observer in a heliocentric reference frame.

"No, it's not."

Yes, it is.

"Dark matter has nothing to do with the claim that the 'galaxies should fall part due to their rotation.' Really, where do you get this stupid idea?"

I bought it at the Stupid Idea Store. It was on sale for a dollar. Do you think I paid too much? Anyway. Fine. Forget the part about galaxies falling apart due to rotating too fast. Assume I was in error regarding the falling apart aspect. I misspoke. I was being completely stupid and totally in error about the galaxies falling apart. I completely miscommunicated what I meant to say. Still does not change the fact that the observed rotation speeds of galaxies measured over a large range from their centers to their outer edges do not fit theoretical predictions, and the explanation put forward for the discrepancy between theory and observation is dark matter. Look up 'galaxy rotation curve' on wikipedia. Plenty of citations at the bottom. But I know, wikipedia is a poor source of information, right? Still doesn't change the fact that observation does not fit fact regarding galaxy rotation, and the explanation put forward is dark matter. So dark matter actually has everything to do with galaxy rotation, if not with galaxies falling apart due to rotation.

"Your assertion that 'neither Newton nor Relativity can explain why galaxies are spinning faster than they should according to either theory.'" [in response to 'citation needed for what?']

Fine. They can explain it. But their explanation is to hypothesize the existence of a form of matter that has only been detected through a discrepancy between theoretical prediction and empirical observation. Dark matter is an ad hoc explanation.

In that spirit, I hereby explain why we are inside a black hole and yet are not crushed to death. The universe is permeated by some sort of physical force or matter that prevents everything from being crushed inside our black hole universe. SOMETHING exists that makes our geocentric, black hole universe habitable. That something MUST exist, because we're here and my theory (for the purposes of this sentence) doesn't work without it. There. If such ad-hoc explanations as dark matter are permissible, then mine is too. X explains how we can live inside a black hole, and X is whatever it needs to be. Matter, energy, a pink elephant, whatever it needs to be. No theory need ever be wrong again. X can always plug the hole.

"Dark Matter falls perfectly within Newtonian mechanics."

Citation needed.

Dark matter MIGHT fall perfectly within Newtonian mechanics if we knew what it was. But since it's never actually been detected other than discrepancy between theory and observation, we have no idea what dark matter and dark energy actually are. It hasn't even yet been determined whether we are detecting the presence of dark matter, or whether we are detecting the presence of errors in standard theories. Right now, the best that can be said about dark matter is that it MUST exist IF standard cosmological theories are valid. But that doesn't mean it DOES exist. It could just mean that standard theories don't work.

"It's because of that fact that we can detect its presence, since it has a definitive gravitational influence on surrounding matter."

It's not a fact that we can detect its presence. It is only a fact that we can detect a discrepancy between theoretical prediction and empirical observation, and dark matter is generally regarded as the culprit.

"Good."

That's your response to my statement 'Possibly we wouldn't have such a reason.' But I made that statement with a huge qualification which you are not addressing.

"No, you are making him right by misrepresenting his position. As long as you keep doing that, I will ignore the rest of your blather, since it's nothing but a dishonest move anyway."

Dude, you misunderstanding his position is not the same thing as me misrepresenting it. Go ahead and ignore my blather; even IF I am wrong, there's nothing dishonest about it. It's a genuine and honest misunderstanding of Einstein's position. But I'm not wrong.

"There you go."

I made a highly qualified concession, but you're taking it as an unqualified concession. So you're following through on your vow to ignore my alleged blather, but only after you pick out a portion that misrepresents my true position. Now who is being dishonest and misrepresentative?

"Well, I am not the one doing the quote-mining."

Neither am I. At the beginning of our discussion, I facetiously labeled myself as a quote-miner. But absolutely no quote I've used in this discussion has been taken out of context or misinterpreted.

If I cite a source but don't offer a quote from the source, you'll demand that I point to a particular section that supports my position. If I then offer a quote from the source but it supports my position, you'll accuse me of being a quote-miner and taking the quote out of context and demand that I give a larger quote to offer a larger clue to the context. But with the quotes I HAVE offered, no matter how much I expand the quote, it won't alter the quote's support for my position, since I'm not taking it out of context.

"You failed to point out any inconsistency."

Fine. Whether there was or was not an inconsistency is not

important to me, so for the purposes of this discussion, there was no inconsistency.

"It's you who has no idea what a reference frame really is."

Is not.

[no further responses as yet]

NOTE: when I say "no further responses as yet" in these debates, I probably should add "as far as I know." I'm only aware of responses that show up in my Google+ notifications, so if someone doesn't actually reply to one of my posts, but instead just posts a reply to the comments thread as a whole, it won't show up in my notifications. So "as far as I know" is the caveat regarding no further responses.

Debate Seven

Scott Reeves vs. MomoTheBellyDancer
Round II

Comments on the YouTube video "Gravitational Wave Hoax - LIGO fake blind injection discovery" by Russ Brown https://www.youtube.com/watch?v=oed1Uqx9tQE&feature=youtu.be

<u>**Scott Reeves (in response to comments by Indra) wrote**</u>:

"Hopefully something better will come up soon in this century."

It has. Geocentrism has returned from exile, and is going to gradually sweep Einstein's ideas out the door like the nasty dust bunnies that they are.

<u>**MomoTheBellyDancer wrote:**</u>

" Geocentrism has returned from exile"

No, it's dead, and has been for a long while now. The only people who keep beating that dead horse are kooks, and they usually do that for religious reasons.

Scott Reeves wrote:

It's clearly not dead. It's currently enjoying a resurgence. And I've never advocated absolute Geocentrism for religious reasons. I advocate it because Relativity is a faulty and untested theory. It is pseudoscience, and the alternative to it is absolute Geocentrism.

MomoTheBellyDancer wrote:

"It's currently enjoying a resurgence."

Only among the scientifically illiterate.

Scott Reeves wrote:

"Only among the scientifically illiterate."

There's nothing scientifically illiterate about advocating geocentrism. Einstein himself advocated relativistic geocentrism. Was he scientifically illiterate? I'm fully aware that Relativity forbids absolute Geocentrism, yet I advocate it anyway because I'm not a relativist. There is nothing scientifically illiterate about my position. I can and have defended it against you, so if I'm scientifically illiterate yet can hold my own in a debate with you, what does that say about your own scientific literacy?

MomoTheBellyDancer wrote:

"There's nothing scientifically illiterate about advocating

geocentrism. "

Geocentrism can only be arrived at by ignoring vast swaths of scientific knowledge.

" Einstein himself advocated relativistic geocentrism. "

No, he didn't. This is a lie, perpetuated by the scientifically illiterate.

"I'm fully aware that Relativity forbids absolute Geocentrism,"

Then how could Einstein have advocated "relativistic geocentrism?" You're not making any sense.

"I advocate it anyway because I'm not a relativist."

What the heck is "relativist?" Relativity is accepted because of the evidence, not because people "believe" in it.

"There is nothing scientifically illiterate about my position."

Your position is a prime example of scientific illiteracy, surpassed only by flat-earth idiocy.

" I can and have defended it against you, so if I'm scientifically illiterate yet can hold my own in a debate with you,,"

You got shot down at every turn, so your self-assurance borders on the delusional. You're a prime example of the Dunning-Kruger effect.

<u>Scott Reeves wrote</u>:

"Geocentrism can only be arrived at by ignoring vast swaths of scientific knowledge."

Break these vast swaths down into an itemized list to which I can respond. As it stands, your claim is too nebulous and ill-defined to allow a sensible response. I need to know your itemized composition of these vast swaths that I and other geocentrists are allegedly ignoring.

"No, he didn't. This is a lie [that Einstein advocated relativistic geocentrism], perpetuated by the scientifically illiterate."

Yes, he did, and no, isn't. I would insert numerous quotes from Einstein where he specifically discusses the geocentric reference frame, but you'd just accuse me of quote mining. Because obviously Einstein would never have said even the slightest word favorable to geocentrism.

And just to clarify your position: you equate scientific illiteracy with rejection of a theory favored by a majority of scientists?

"Then how could Einstein have advocated 'relativistic geocentrism?' You're not making any sense."

He advocated relativistic geocentrism because he couldn't advocate absolute Geocentrism, since Relativity forbids an absolute reference frame but MUST allow a geocentric reference frame. I actually am making sense, but you need to understand the difference between relativistic reference frames and absolute reference frames in order to realize that I'm making sense.

"What the heck is 'relativist?'"

http://dictionary.reference.com/browse/relativist?

"Relativity is accepted because of the evidence, not because people "believe" in it."

Incorrect. Relativity isn't accepted because of the evidence. People who accept relativity do so because they believe in a relativistic interpretation of the evidence. Just as people who accept absolute Geocentrism do so because they believe in a non-relativistic interpretation of the evidence.

"Your position is a prime example of scientific illiteracy, surpassed only by flat-earth idiocy."

Says a relativist who has to ask what a relativist is. Anyway, you've got the kook scale wrong. It goes absolute Geocentrism, relativity, flat-earth.

"You got shot down at every turn, so your self-assurance borders on the delusional. You're a prime example of the Dunning-Kruger effect."

Which is exactly what a person who doesn't understand either sort of geocentrism enough to mount an effective argument against absolute Geocentrism would erroneously think, being themselves a prime example of the Dunning-Kruger effect.

It's not your fault, though, and you are not alone. You just have been so thoroughly indoctrinated into the Einsteinian worldview that you are literally unable to even entertain the possibility that a valid alternative interpretation of the evidence might exist that fundamentally violates your worldview, and cognitive dissonance renders any argument put forward in favor of such an alternative interpretation as nonsensical to you.

To reduce the dissonance, you believe that anyone who argues against your theory, no matter how literate they are in your theory, must be a prime example of the Dunning-Kruger effect. They can't be anything but, because if they actually do understand your theory yet still reject it, then your view of reality will be shattered. You literally could not handle the truth of any opposing world view. Your brain would implode. Or perhaps it would explode. The only way to determine which it would be is to convert you to absolute Geocentrism and observe the result. When and if that ever happens, we will write a scientific paper on the type of brain trauma that occurs when converting to absolute Geocentrism. I myself did not experience any such trauma when converting because, of course, my brain was already damaged to begin with due to prolonged exposure to relativity.

Modern mainstream scientists now occupy the same inflexible

position that the Church did back in the days of Copernicus, except that modern scientists are even more inflexible than the Church was, because they believe that reason, logic and intelligence guide them much more surely and infallibly than God and the Bible supposedly guided the Church.

MomoTheBellyDancer wrote:

"Break these vast swaths down into an itemized list to which I can respond."

Translation: "I'm too lazy to do any actual research and want you to do it for me." Sorry, but I have better things to do than to aid your hubris. If you want to go against established science *you* should provide a list of observations you think support *your* model. And remember: extraordinary claims require extraordinary evidence.

"Yes, he did"

No, he didn't. Stop lying.

"you equate scientific illiteracy with rejection of a theory favored by a majority of scientists?"

I do, if that rejection is based on nothing but scientific misconceptions.

"He advocated relativistic geocentrism because he couldn't advocate absolute Geocentrism"

Bullshit.

"http://dictionary.reference.com/browse/relativist"

Argumentum ad Dictatoriam. Fail.

"Relativity isn't accepted because of the evidence. People who accept relativity do so because they believe in a relativistic interpretation of the evidence."

Once again, you show your ignorance of the scientific method. Science is about making models of reality. Einstein's model is accepted because it explains observations to a high degree of accuracy, and because it offers predictions that have been shown to be correct (gravitational waves, for instance). It *works*.

"Just as people who accept absolute Geocentrism do so because they believe in a non-relativistic interpretation of the evidence."

All they do is reject a highly successful model just because it doesn't vibe with their preconceived notions. They offer no model that explains the same observations equally well, and with the same amount (or even less) assumptions.

"Says a relativist who has to ask what a relativist is."

Just putting "-ist" after a word means jack shit.

"Anyway, you've got the kook scale wrong. It goes absolute

Geocentrism, relativity, flat-earth."

Geocentrism is kookery, since it is nothing but science denial.

"Which is exactly what a person who doesn't understand either sort of geocentrism enough to mount an effective argument against absolute Geocentrism"

The argument is that it is contradicted by observations, which is the death knell for *any* hypothesis.

"You just have been so thoroughly indoctrinated into the Einsteinian worldview"

And in the end, all you kooks ever offer is conspiracy nuttery. No surprise here.

<u>Scott Reeves wrote:</u>

+MomoTheBellyDancer

"Translation: 'I'm too lazy to do any actual research and want you to do it for me.' Sorry, but I have better things to do than to aid your hubris. If you want to go against established science you should provide a list of observations you think support your model. And remember: extraordinary claims require extraordinary evidence."

It had nothing to do with being lazy. You accused me and all geocentrists of ignoring vast swaths of scientific knowledge, and I wanted an itemized list of exactly what evidence you allege that we're ignoring. Despite your allegations of my scientific illiteracy, I already know the evidence that allegedly supports Relativity, as well as the evidence that supports absolute Geocentrism, so I don't need you to dig up evidence for or against either my position or relativity's. I want to know exactly what's on your list of things you believe are being ignored so that I can determine whether or not I'm ignoring any of it. As it stands, your nebulous claim that I'm ignoring vast swaths of scientific knowledge only deserves the response, "No, I'm not." Your accusation wouldn't survive in a court case because it lacks specificity. You don't make vague accusations against someone and then further accuse them of being lazy if they demand that you provide them with a list of specific accusations.

"No, he didn't. Stop lying."

Yes, he did. And I'm not lying. Me lying is not the same as you refusing to accept that Einstein demonstrably said certain things.

"I do, if that rejection is based on nothing but scientific misconceptions."

If.

"Bullshit."

So Einstein COULD advocate absolute Geocentrism?

"Argumentum ad Dictatoriam. Fail."

The argument for dictatorship? I have no idea what that means in the context of our discussion.

You asked me what a relativist was. I supplied the answer. Success. If you were looking for something other than a valid dictionary definition, you should have specified that. Besides: translation: "I'm too lazy to do any actual research and want you to do it for me." Do your own research on what a relativist is.

"Science is about making models of reality. Einstein's model is accepted because it explains observations to a high degree of accuracy,"

So explain relativity's geocentric model, and give examples of how its explanatory power is superior to the absolute Geocentric model.

"and because it offers predictions that have been shown to be correct (gravitational waves, for instance). It works."

You mean that recent one-off event that has yet to be replicated as per the scientific method? That one-off event that was created by the movement of massive objects rather than by gravity itself? Anyway, I don't have a problem with gravitational waves. It would be foolish to think that mass moving through space wouldn't generate ripples in the ether, whether it's the classic ether or Einstein's version of it. Where I disagree with Relativity is its rejection of an absolute reference frame, which has no bearing on the existence of gravitational waves.

"All they do is reject a highly successful model just because it doesn't vibe with their preconceived notions."

I'm an absolute Geocentrist, and I don't reject a highly successful model that doesn't vibe with my preconceived notions. I reject an untested theory called Relativity on the grounds that it is pseudoscience.

"They offer no model that explains the same observations equally well, and with the same amount (or even less) assumptions."

Again, you are the one claiming the absolute Geocentric model doesn't explain the same observations equally well. Provide a list of these observations which you allege are not explained equally well. The same with your allegation about the 'same amount (or even less) assumptions.' Provide a list of the assumptions of both models which shows that the absolute Geocentric model has more assumptions than the relativistic geocentric model.

"Just putting "-ist" after a word means jack shit."

Hey, I didn't invent the English language. But "-ist" means a whole lot more than jack shit. It is a hugely important suffix. A person who adheres to relativity is a relativist. A person who adheres to geocentrism is a geocentrist. A person who practices optometry is an optometrist. Gynecologist. Jurist. Scientist. Flutist. Cellist. Optimist. Pessimist.You asked me what a relativist was, and I supplied the answer. The real issue is, why did you have to ask what a relativist is if you are scientifically literate as well as literate in the English language?

"Geocentrism is kookery, since it is nothing but science denial."

Geocentrists don't deny science. If anything, relativists deny science when they have an untested hypothesis that there is a universe beyond the boundary of our observable universe (all points will look like they're at the center), a larger, UNOBSERVABLE universe which Relativity requires in order to get us out of literally being at the center of the whole universe, which we empirically are if our observable universe is all the universe there is, yet they present their theory to the public as the most successful theory in the history of science. But since anything outside our observable universe is UNOBSERVABLE, then it is beyond the scope of rational scientific inquiry, hence Relativity is an unscientific theory.

"The argument is that it is contradicted by observations, which is the death knell for any hypothesis."

That may be the argument you're trying to mount, but you don't seem to be knowledgeable of the position of absolute Geocentrists, and you don't seem to understand the relativistic geocentric model either. For Relativity to be a valid theory, ANY observation you could possibly throw onto your list of observations that allegedly contradict geocentrism MUST be explicable solely from the viewpoint of an observer in the geocentric reference frame, without involving an observer in an external reference frame. Meaning that relativity's geocentric observer MUST have an entirely geocentric explanation for everything he observes. If you argue that a certain observation contradicts absolute Geocentrism, that same observation still has to be geocentrically explicable by Relativity's geocentric observer, so by arguing against absolute Geocentrism, you have to be careful that you don't shoot yourself in the foot and deny relativistic geocentrism as well. Based on all you've said to me, you don't seem to realize this, which is why, where geocentrism is concerned, you are a prime example of the Dunning-Kruger effect.

"And in the end, all you kooks ever offer is conspiracy nuttery. No surprise here."

I offered no conspiracy nuttery. I said you had been thoroughly indoctrinated into an Einsteinian worldview. But I mentioned no conspiracy to indoctrinate people into Relativity, and I didn't mean to imply one. Such indoctrination isn't due to a conspiracy, any more than religious indoctrination is due to a conspiracy. Or perhaps you believe religious indoctrination IS a conspiracy, in which case you yourself are a conspiracy nutter. Indoctrination is possible without a conspiracy [citation needed? Too bad]. And here, I'm going to say conspiracy one more time.

MomoTheBellyDancer wrote:

"I wanted an itemized list of exactly what evidence you allege that we're ignoring."

You have to ignore practically *everything* we learned in last 500 years with regard to astronomy, physics, geology and optics. That's a lot of knowledge to itemize.

Again, if you think you have a case against all those centuries of hard-won scientific insights, then go for it.

Scott Reeves wrote:

"You have to ignore practically *everything* we learned in last 500 years with regard to astronomy, physics, geology and optics. That's a lot of knowledge to itemize."

Ignoring observations and ignoring what other people have learned are two different things. For example, for 500 years people have gathered observations in astronomy and physics, and from those observations, some people learned that we live in a relativistic universe. Other people learned from those same observations that we live in an absolutely Geocentric universe. The mere fact that a group of people learned something doesn't automatically mean they learned the truth. I'm not ignoring any observations, I'm ignoring what relativists think they have learned from those observations, because they didn't learn the truth.

MomoTheBellyDancer wrote:

"Ignoring observations and ignoring what other people have learned are two different things."

It is the same thing when those things that have been learned are *based* on those observations. You;re basically calling every scientist out there a liar for no good reason whatsoever.

"for 500 years people have gathered observations in astronomy and physics, and from those observations, some people learned that we live in a relativistic universe"

What? No. Einstein came up with the theory of Special Relativity in 1905. What the heck are you even talking about?

"Other people learned from those same observations that we live in an absolutely Geocentric universe."

Only if you distory those observations as to be practically unrecognizable.

" because they didn't learn the truth."

And you have "learned the truth" (whatever that means) ...*how* exactly?

<u>Scott Reeves wrote:</u>

"What? No. Einstein came up with the theory of Special Relativity in 1905. What the heck are you even talking about?"

We were discussing geocentrism, and you claimed that geocentrists "have to ignore practically *everything* we learned in last 500 years with regard to astronomy, physics, geology and optics." I then replied that some people learned from those five hundred years of observations that we live in a relativistic universe, and some learned that we live in an absolute Geocentric universe. I know Special

Relativity was foisted upon the world in 1905. I also know General Relativity was foisted upon the world ten years later. What does that have to do with anything? Is Special Relativity not one of those things you say we learned in the last 500 years? What the heck are YOU even talking about? I sense a miscommunication/misunderstanding here.

"It is the same thing when those things that have been learned are *based* on those observations."

No, it's not the same thing. "Based on" being the key. You're just proving my point. You're making a distinction between things that have been learned and observations. If Steve learned X based on Z, and Mortimer learned Y based on Z, and Steve ignores what Mortimer learned, it's not the same as Steve ignoring Z. Steve is ignoring Y based on Z.

But Steve and Mortimer each can learn both X and Y. And if X and Y are incompatible, then they can debate whether X or Y is the true picture of reality based on Z. But if Steve is the only one that learned both X and Y, then Mortimer is at a disadvantage.

And anyway, "ignoring" what we've learned is not the correct word. "Rejecting" is more applicable.

"You;re basically calling every scientist out there a liar for no good reason whatsoever."

No. Disagreement with scientists' interpretation of observations is not the same thing as calling them liars.

"Only if you distory those observations as to be practically unrecognizable."

They're not being distorted. They only appear distorted to you because of cognitive dissonance. Realizing that we live in an absolute Geocentric universe is like a Matrix-level revelation, and some people can't handle it.

"And you have 'learned the truth' (whatever that means)"

How about I rephrase what I said and say that I learned both models of the universe (the relativistic and the non-relativistic) and rejected the relativistic model as not being representative of reality.

"... *how* exactly?"

Same as you, I assume. Study. But I have a leg up on you, because I studied competing models that are based upon the same observations, and the possibility that Earth might literally be at the center of the universe doesn't bother me, and it also doesn't bother me if it isn't at the center of the universe. So I'm able to understand and consider both models rationally and dispassionately and come to the conclusion that the evidence currently available to us best supports the absolute Geocentric model. What evidence? Every observation we've gathered in the last 500 years.

MomoTheBellyDancer wrote:

"Is Special Relativity not one of those things you say we learned in the last 500 years?"

Yes, but it's by far not the only reason we know geocentrism is false. We have been knowing this for centuries already.

"rejected the relativistic model as not being a true representation of reality."

Based on nothing but your own ignorance, no less.

" I studied competing models that are based upon the same observations"

No. All the observations lead only to one model of earth's place in the universe, and this model isnot geocentric.

Earthbound wrote (inserted here because it was in thread with Momo):

+Scott Reeves

I've finally stopped laughing -- Shakespeare "taught" us some "500 years ago" that ignorance is a part of the tragical comedy of life. When we need to use the word "taught" considering theories, maybe we're already on the "wrong" track and need to go back to defining what "a theory" is all about. The word "researcher" in French translates to "chercheur", someone who seeks. Most of those in fundamental research repeat over and over that they hope someone will come along and "prove to be wrong" what they and everyone else in their field has believed was "right". Of course they're happy when modern "tools" seem to confirm their theories, but that's not what they really are interested in in the long run. The simple reality of how vast the Universe really is and how amazing our place is here should be enough to send us wishing everyone could appreciate Carl Sagan's message. There are a lot of people out there trying to get it across. When I read such nonsense as some that is posted here, it does motivate me to write about things I do know first hand -- physics isn't one -- and to listen to those whose lives are turned to understanding beyond just opinions, unless it's just a game not to be taken seriously (or to make an "extra buck" off of YouTube clicks)? (lol)

Scott Reeves wrote:

+Earthboud

"Shakespeare "taught" us some "500 years ago" that ignorance is a part of the tragical comedy of life."

Just to be clear, I was not the one who brought up this whole "500 years ago" stuff. That was MomoTheBellyDancer.

"When we need to use the word "taught" considering theories, maybe we're already on the "wrong" track and need to go back to defining what "a theory" is all about."

I don't need to use the word "taught" considering theories, and I never did use it. MomoTheBellyDancer used the word "learn" when he said, "You have to ignore practically everything we learned in last 500 years with regard to astronomy, physics, geology and optics. That's a lot of knowledge to itemize." He defined that part of the discussion in terms of the word "learn," and I simply kept to his definition in my subsequent comments.

"Most of those in fundamental research repeat over and over that they hope someone will come along and 'prove to be wrong' what they and everyone else in their field has believed was 'right'"

That's only true to a certain extent. They don't want to be proven TOO wrong, because it seems that as "chercheur" they aren't willing to go too far off course in their search. For example, you and your physicist companion, from your constant references to laughter, do nothing but laugh at anyone who advocates a view of the universe that is too radical a departure from accepted dogma. If they were true "chercheur" then they would seek out and be open to seriously considering ANY idea, no matter how seemingly kooky, because it might hold a grain of truth that could help them in their search for scientific truth. They would even be willing to circle back and revisit ideas that had been discarded centuries earlier. As "chercheur" they confine themselves within narrow boundaries, and so aren't willing to journey far in their search. If you want to go to a certain destination, you might never get there if you insist upon only traveling certain roads.

"The simple reality of how vast the Universe really is and how amazing our place is here should be enough to send us wishing everyone could appreciate Carl Sagan's message. There are a lot of people out there trying to get it across."

You seem to be implying that the Universe wouldn't be just as vast and amazing if we were at the center of it. Why wouldn't it be? Why wouldn't an absolute Geocentric universe be equally as amazing?

MomoTheBellyDancer wrote (to Earthbound):

" that ignorance is a part of the tragical comedy of life. "

And Martin Luther King once said: "Nothing in the world is more dangerous than sincere ignorance and conscientious stupidity." Scott exemplifies those perfectly.

Earthbound wrote:

+MomoTheBellyDancer

Well, Momo, if nothing else comes out of all this -- mostly, it gives me some notion of "popular culture" since I'm outside of the country. I'm relieved to know that there is still a little sanity to balance things out! (lol) It's also encouraged me to get around to setting up a YouTube channel -- I downloaded the necessary program this am. Though some of the videos will also be in French (I'll try to find ones with English subtitles), hopefully those who are open to more serious reflection will find something of interest there.

Scott Reeves wrote:

+MomoTheBellyDancer

"Yes, but it's by far not the only reason we know geocentrism is false. We have been knowing this for centuries already."

How have we "been knowing" that geocentrism is false? Focault's pendulum? Only shows that there is relative motion between the Earth and the fixed stars. Moons of Jupiter? Only shows that individually, not everything orbits the Earth; does not speak to whether the universe as a whole revolves around the Earth. Phases of Venus? Only shows that the geocentric model needed to be modified so that Venus orbits the sun rather than the Earth as it did in the Ptolemaic model. Does not speak to whether the universe as a whole revolves around the Earth. Stellar parallax and stellar aberration? If you actually delved into modern geocentrism, you would know that geocentrism explains them both. More massive objects don't orbit less massive objects? Not a problem; the sun isn't orbiting the Earth, it's orbiting the barycenter of the entire universe. Stars spinning faster than light in a geocentric universe? Not an argument against geocentrism, according to Einstein and various other physicists. Geostationary and geosynchronous satellites? There are geocentric explanations for those as well, which you would know if you had actually studied the geocentric model. GPS? Geocentrism predicts the same needed corrections for the GPS system, using the exact same reasoning as Relativity. Interferometer experiments? Support for geocentrism.

I'm like the Energizer bunny. I could keep going on and on, but there's no need. Any observation you could possibly throw out can be explained by geocentrism. I've said this multiple times before, but you have yet to respond to it (at least I think I said it to you at one point, sorry if I'm mistaken about that). If you think there's an observation that can't be explained by a geocentric observer, then Relativity is invalidated. We have not "been knowing" that geocentrism is false for centuries (Well, we allegedly did, but then Relativity came along). Geocentism is not false, not according to Einstein, and not according to absolute Geocentrists. Geocentrism is unquestionably true. The only question is whether it's only true from the perspective of a geocentric observer, or whether it is absolutely true and Relativity is false. I know you believe I'm lying about that, but his words are there for posterity, regardless of whether you accept them or not. This is something that you and other people arguing against any form of geocentrism just do not seem to understand. Einstein himself understood that geocentrism has to be just as correct as any other viewpoint, else Relativity is false. Relativity merely forbids anyone from claiming that the geocentric reference frame is in any way preferred over other reference frames. If you are anti-geocentrist period, then you are implicitly anti-Relativity. Someone who actually understands Relativity knows that he/she can only ever be an anti-ABSOLUTE Geocentrist. Anyone who does not understand this is ignorant of Relativity at a fundamental level.

"Based on nothing but your own ignorance, no less."

Argumentum ad hominem..

"No. All the observations lead only to one model of earth's place in the universe, and this model isnot geocentric."

No. According to Relativity, all observations lead to two equally valid models of Earth's place in the universe: a geocentric one, and a non-geocentric one. The fact that you would make such a statement shows that you do not understand this aspect of Relativity. Relativity denies that the geocentric model represents a preferred reference frame. Absolute Geocentrists say that it does represent a preferred reference frame.

Earlier in our back and forth, in comments on another video, you finally admitted, "I am not saying that you can't have a geocentric reference frame." If there is a geocentric reference frame, then an observer within that reference frame must have a model for how the universe works from his geocentric perspective, and that model cannot appeal to an outside reference frame for explanations. If it has to make such an appeal, then the geocentric reference frame is an inferior reference frame, and Relativity is invalidated. Therefore, if you believe that you can put forward some observation that disproves geocentrism, then you implicitly believe that Relativity is an invalid theory.

You as a relativist MUST be able to assume the role of a geocentric observer and say, "Okay, Earth is at rest and everything else in the universe is in motion. Given this situation, what is my explanation for how the universe works?" That explanation is your geocentric model. If you are unwilling to assume such a role, then you are an anti-relativist. If you deny that you must be able to assume such a role, then you do not truly understand Relativity. If you cannot produce such geocentric explanations, then Relativity is false. This is not my opinion; this is fact regarding Relativity. So bring forth your aforementioned observations by which we absolutely know that geocentrism is false. You'll be disproving Relativity, but at least you'll shut me up regarding geocentrism.

"And Martin Luther King once said: 'Nothing in the world is more dangerous than sincere ignorance and conscientious stupidity.' Scott exemplifies those perfectly."

And Socrates once said, "When the debate is lost, slander becomes the tool of the loser."

MomoTheBellyDancer wrote:

"Only shows that there is relative motion between the Earth and the fixed stars."

"Only shows that individually, not everything orbits the Earth; does not speak to whether the universe as a whole revolves around the Earth."

"Only shows that the geocentric model needed to be modified so that Venus orbits the sun rather than the Earth as it did in the Ptolemaic model."

"Not a problem; the sun isn't orbiting the Earth, it's orbiting the barycenter of the entire universe."

"GPS? Geocentrism predicts the same needed corrections for the GPS system, using the exact same reasoning as Relativity. Interferometer experiments? Support for geocentrism."

The problem is that, under geocentrism, you need to insert massive amounts of variables the system to explain any of these. The model of a moving earth explains all the above in *one fell swoop,* along with many others (like the retrograde motion of planets, the slight size in angular size between perihelion and aphelion, the seasons, the precession of the equinox). Occam's Razor compels us to go with the model with the least amount of assumptions. Under geocentrism, you have to pile assumption on assumption in an effort to make the model work, and then it still fails. The geocentric model is like a boat full of holes, where fixing one hole leads to the creation of ten others.

"Argumentum ad hominem."

Nope.

"According to Relativity, all observations lead to two equally valid models of Earth's place in the universe"

Wrong. But I already explained that.

"I am not saying that you can't have a geocentric reference frame."

Oh, you can have your belly button as a reference frame, but that doesn't mean the universe actually revolves around your belly button.

"You as a relativist"

No such thing exists, just as there are no evolutionists, gravitationists or plate tectonicists. Putting -"ism" behind words is just a cheap poly to make them sound like beliefs.

"When the debate is lost, slander becomes the tool of the loser"

Reducing everything to an "ism" *is* a form of slander.

Scott Reeves wrote:

"The problem is that, under geocentrism, you need to insert massive amounts of variables the system to explain any of these."

You've said that several times, but you have yet to list even a few of the assumptions of either model, let alone a complete list of assumptions of both models which shows that the geocentric model indeed has more assumptions. I suspect that a lot of the assumptions you attribute to the geocentric model are actually just statements that the geocentric observer has to assume that he's in a geocentric reference frame, which no duh.

But regardless that, Relativity has to make the biggest assumption of them all: that there is a larger universe beyond the universe as observed from Earth. If that larger universe does not exist, then observation shows that we are literally at the center of the entire universe. The problem for Relativity is that anything beyond our observable universe is UNOBSERVABLE, and thus beyond the scope of rational scientific inquiry. In other words, Relativity depends upon the existence of something that cannot be scientifically proven to exist.

But let's pretend that you're correct, and that "The problem is that, under geocentrism, you need to insert massive amounts of variables the system to explain any of these." If it's actually a problem, then it's a problem for Relativity as well, because Relativity's geocentric observer still has to be able to explain everything solely from a geocentric perspective. If he has to insert massive amounts of variables to make his geocentric model work, then that's what he has to do to make Relativity a valid theory. You're not helping Relativity's case by asserting the need for these massive amounts of variables.

"The model of a moving earth explains all the above in *one fell swoop*, along with many others (like the retrograde motion of planets, the slight size in angular size between perihelion and aphelion, the seasons, the precession of the equinox)."

So does geocentrism, whether it's relativistic or absolute.

"Occam's Razor compels us to go with the model with the least amount of assumptions."

Occam's Razor is a philosophical desire for simplicity, not a physical law that governs the universe. Reality compels us to accept whatever sort of universe we actually live in. And Relativity compels us to go with the model of whichever observer we're pretending to be.

"Under geocentrism, you have to pile assumption on assumption in an effort to make the model work, and then it still fails. The geocentric model is like a boat full of holes, where fixing one hole leads to the creation of ten others."

"Boo-hoo," whined the relativist. "My geocentric reference frame doesn't work." Then neither does your theory. If the geocentric model is invalid, then we live in a non-geocentric, non-relativistic universe. I'm fine with that.

"Wrong. But I already explained that."

You explained your position on the subject. But your position is the position of someone who misunderstands something foundational to Relativity.

"Oh, you can have your belly button as a reference frame, but that doesn't mean the universe actually revolves around your belly button."

It does if your belly button is coincident with the barycenter of the entire universe.

"No such thing exists, just as there are no evolutionists, gravitationists or plate tectonicists."

Evolutionist: http://dictionary.reference.com/browse/evolutionist
Relativist: http://dictionary.reference.com/browse/relativist

Next you're going to be telling me there are no gynecologists, no biologists, no optometrists, no proctologists, no dentists, no flutists, no cellists, no geologists, no planetologists, no astrologists, no ecologists, no vulcanologists etc. I'm just curious: is English your native language?

It would be nice if there were no relativists, but sadly, there are.

"Reducing everything to an "ism" *is* a form of slander."

Wait a minute. Are you telling me that whoever coined the term "geocentrism" meant it as slander or libel? And that's why you've been using it? And I've been going along with it? That really chaps my ass. I hereby demand that all adherents to Einstein's Relativism cease and desist referring to geocentricity as geocentrism, or I shall sue for defamation on behalf of all geocentrists everywhere (which includes anyone who truly understands Einstein's Relativism).

MomoTheBellyDancer wrote:

"You've said that several times, but you have yet to list even a few of the assumptions of either model, let alone a complete list of assumptions of both models which shows that the geocentric model indeed has more assumptions."

Assumptions for the moving earth model (which are supported by observations): The earth rotates at a tilt, orbits the sun with the other planets, and the sun rotates around the center of the galaxy. With this you can solve a vast majority of the issues mentioned in one go.

For the geocentric model you will have to add an assumption for every observation you make, which quickly runs into the dozens, if not *hundreds*. You need to add assumptions fot the seasons, the retrograde motion of planets, the precession of the equinox, the phases and transits of Venus and Mercury, the existence of the Milky Way, the Coriolis effect, Foucault's pendulum, earthquakes, earth magnetism, and so on, and so forth. And the more assumptions you add, the less it becomes a cohesive whole.

"Relativity has to make the biggest assumption of them all: that there is a larger universe beyond the universe as observed from Earth."

No, it doesn't. Relativity is largely about the motion of bodies. The observation that there is likely a bigger universe outside what we observe comes from astronomy and cosmology.

"If that larger universe does not exist, then observation shows that we are literally at the center of the entire universe."

Nonsense. The part of the universe we *can* observe already indicates that we are definitely not at the center of it.

"The problem for Relativity is that anything beyond our observable universe is UNOBSERVABLE"

It still leaves an affect on *our* part of the universe, that we can observe and measure. The cosmic background radiation is one such example.

"In other words, Relativity depends upon the existence of something that cannot be scientifically proven to exist."

It becomes more and more obvious that you have absolutely no idea what relativity even entails.

"If that's actually a problem, then it's a problem for Relativity as well, because Relativity's geocentric observer still has to be able to explain everything solely from a geocentric perspective."

Only if he adds massive amounts of *other* assumptions as well. You still fail to realize that the moving-earth model does *not* depend on relativity whatsoever.

"So does geocentrism, whether it's relativistic or absolute."

No, it doesn't. For instance, by what mechanism would the sun, Moon and stars rotate around the earth as they do at those speeds?

"Occam's Razor is a philosophical desire for simplicity, not a physical law that governs the universe."

Why should we use a model with added assumptions, if a simpler model works just as well, if not *better*?

""Boo-hoo," whined the relativist"

There is no such thing as a relativist.

"'My geocentric model doesn't work.' Then neither does your theory."

There is massive amounts of evidence for Einstein's theories. There is no evidence for geocentrism.

"If Relativity's geocentric model is invalid, then we live in a non-geocentric, non-relativistic universe. I'm fine with that."

Except for the fact that those two are unrelated models. You're comparing apples to jumbo jets.

"You explained your position on the subject. But your position is the position of someone who misunderstands something foundational to Relativity."

This is rich, coming from someone who has shown time and time again not to grasp relativity *at all.*

"It does if your belly button is coincident with the barycenter of the entire universe."

The universe has no barycenter, so nothing can coincide with it.

"Evolutionist:http://dictionary.reference.com/browse/evolutionist"
"Relativist:http://dictionary.reference.com/browse/relativist"

Argumentum ad Dictatoriam is *still* fail.

"Next you're going to be telling me there are no gynecologists, no biologists, no optometrists, no proctologists, no dentists, no flutists, no cellists, no geologists, no planetologists, no astrologists, no ecologists, no vulcanologists etc."

So are you saying that there is something as gynecologism? Biologism? Geologism? Vulcanologism? Are you claiming all those positions are mere opinions?

You see, most of those words are derived from words that end in "-logy." Your grasp of language appears to be as piss-poor as your grasp of science. No surprise there.

"I'm just curious: is English your native language?"

Wow. The irony.

"It would be nice if there were no relativists, but sadly, there are."

Yet there aren't. There are scientists who accept the massive amounts of evidence for Einstein's theories. That doesn't reduce their position to an "ism."

"Wait a minute. Are you telling me that whoever coined the term "geocentrism" meant it as slander or libel?"

"Ism" is generally not used for scientific theories. This is because they are not mere opinions. They are models that comprehensive explain observations in reality.

Again, fix your lack of knowledge of science. Right now you're just embarrassing yourself.

<u>Scott Reeves wrote:</u>

"Assumptions for the moving earth model (which are supported by observations): The earth rotates at a tilt, orbits the sun with the other planets, and the sun rotates around the center of the galaxy. With this you can solve a vast majority of the issues mentioned in one go."

You are correct. Those are all assumptions. Not empirically proven facts. Made by a non-geocentric observer.

"For the geocentric model you will have to add an assumption for every observation you make, which quickly runs into the dozens, if not *hundreds*."

So you're saying you really have no idea how many or how few assumptions the geocentric model allegedly has to make.

"You need to add assumptions fot the seasons,"

What assumptions are those?

"the retrograde motion of planets,"
What's the assumption? I suspect it's not an assumption, but just a case of you going with an obsolete geocentric model.

"the precession of the equinox,"

What's the assumption?

"the phases and transits of Venus and Mercury,"

No assumptions necessary. You're going with an obsolete geocentric model where Mercury and Venus orbit the Earth rather than the sun.

"the existence of the Milky Way,"

You've actually come up with one I don't think I've ever heard before. I know the one about the sun moving through the Milky Way, but what's the assumption about the existence of the Milky Way?

"the Coriolis effect,"

What assumption? That it's a real force generated by the universe rotating around the Earth? If that's the assumption to which you're referring, the non-geocentric model makes the assumption that it's a fictitious force.

"Foucault's pendulum,"

Coriolis force from the rotating universe rather than the rotating Earth. This is the same assumption you listed earlier.

"earthquakes,"

What the assumption? That earthquakes shift or tilt the Earth's axis, and in a geocentric model this means the universe has to instantly shift instead? But earthquakes don't shift Earth's axis. They shift all the stuff on the Earth surface relative to the axis in both the geocentric and non-geocentric view, so it only appears to people on Earth's surface that the axis has shifted. Observers elsewhere in the universe looking at the Earth would not notice a change in the Earth's axial tilt. And the angular momentum doesn't change. And axis in the geocentric model, of course, means the universe's axis of rotation.

"earth magnetism,"

What the assumption?

"and so on, and so forth. And the more assumptions you add, the less it becomes a cohesive whole."

Need to see all your alleged assumptions to determine whether they are actually assumptions that aren't just variations of "You have to assume that the Earth is stationary at the center of the universe" which is basically what all the assumptions you've listed boil down to. Which is equal to the non-geocentric assumption of "You have to assume that the Earth is moving and not at the center of the universe." Anyway, if these are truly problematic assumptions, then Relativity's geocentric model has a lot of problems, and you relativists need to figure it out to save Relativity.

"Relativity is largely about the motion of bodies."

Incorrect. It's largely about the relativity of motion between bodies. There's a huge difference. It's about allowing each observer to correctly claim that he's at rest and the other observer is in motion, be it constant motion or accelerated motion.

"The observation that there is likely a bigger universe outside what we observe comes from astronomy and cosmology."

"Hypothesis," not "observation." And it comes from both. Astronomical observation shows that everything is generally receding from Earth (red shift). This implies we're at the center of the universe. But that's okay, say modern scientists, because every point will see itself as the center of the universe. This hypothesis is only true if an observer at the edge of Earth's observable universe (our horizon, as you probably call it) can see something beyond the edge (or horizon). If there's nothing beyond the edge (or horizon), then Earth is literally at the center of the universe, just as observations show. And if we're literally at the center of the universe, then Earth is in a preferred, unique position, establishing an absolute reference frame, rendering Relativity invalid. Thus, Relativity implicitly depends upon the existence of a larger universe beyond our observable universe, so that it can say we're not at the center of that larger universe. Since the hypothesis that this larger universe exists is untested, and Relativity is presented to the public as scientific fact, Relativity is a pseudo-scientific theory.

"Nonsense. The part of the universe we *can* observe already indicates that we are definitely not at the center of it."

Nonsense. The position of mainstream scientists is that Earth is observably at the center of its observable universe. That's part and parcel of the Big Bang theory.

"It [the unobservable universe] still leaves an affect on *our* part of the universe, that we can observe and measure. The cosmic background radiation is one such example."

No. Do some research before making unsupportable claims.

"Only if he adds massive amounts of *other* assumptions as well."

If that's what he has to do to make his model work, then that's what he has to do. Because if he isn't able to make his model work, then Relativity is an invalid theory.

"You still fail to realize that the moving-earth model does *not* depend on relativity whatsoever."

You are correct. I still fail to realize that, because the moving-earth model does depend on Relativity. If you claim that it doesn't, let's step into the Wayback Machine and travel back to 1887. Provide a non-relativistic, non-geocentric explanation for the Michelson-Morley experiment. One that wasn't already tried and rejected before Einstein foisted Relativity upon an unsuspecting world.

"No, it doesn't. For instance, by what mechanism would the sun,

Moon and stars rotate around the earth as they do at those speeds?"

Inertia. Gravity. Momentum. Whatever explanation works from a geocentric perspective. How does Relativity's geocentric model explain how they rotate around the Earth as they do?

If you're asking how the sun moon and stars began to rotate around the Earth in the first place, mainstream scientists don't have a workable or widely-agreed-upon explanation of how the universe or life began, so absolute Geocentrists aren't required to have one either. But of course I'm sure you know that most absolute Geocentrists are also religious, and THEY would tell you, "God started it all." But I'm not representative of most absolute Geocentrists, and you won't ever find me appealing to God to explain anything.

"The universe has no barycenter, so nothing can coincide with it."

Your statement makes the assumption that our observable universe has a horizon rather than an edge. In other words, it depends upon the truth of the Copernican hypothesis. Now go ye forth and test that hypothesis. Since the Earth-moon system has a barycenter, and the solar system has a barycenter, and the galaxy has a barycenter, and the local cluster has a barycenter, etc, etc, then the universe as a whole lacks a barycenter only if it is infinitely large, without an edge. "Argumentum ad Dictatoriam is *still* fail."

Second time you've used that phrase with me. "Argument for dictatorship." I think you mean "argumentum ad dictionarium."

But I wasn't making an argument. You asked me what a relativist was, and I gave you the dictionary definition of a relativist. You're the one making the argument that there's no such thing as a relativist. Prove it. If they don't exist, why is there a dictionary entry for them, and why doesn't the dictionary define them as "mythical" or "imaginary" as it does with fairies, unicorns, dragons, and other things that don't actually exist?

"So are you saying that there is something as gynecologism? Biologism? Geologism? Vulcanologism? Are you claiming all those positions are mere opinions?"

I'm not saying that at all. Are YOU saying that there are no such things as biologists, or geologists, or gynecologist, or guitarists, or scientists, etc, etc? Tell that to all the people who profess to be those things, and all the companies on the job websites who are hiring biologists, geologists, gynecologists, etc.

"Why should we use a model with added assumptions, if a simpler model works just as well, if not *better*?"

You go with whatever model represents reality, regardless of complexity. I'm already on the record somewhere saying that I don't deny the existence of multiple REFERENCE FRAMES, or that some of them are more convenient for calculations than others. You use which reference frame is most convenient for what you're trying to do. For example, sending a manned mission to Mars, you would start off using a geocentric reference frame, switch to a heliocentric reference frame, and finish with a Mars-centric reference frame. I just deny that all reference frames are equivalent, and so I go with the model of the universe where they aren't all equivalent. If that model is complex, so be it.

"There is no such thing as a relativist."

Proof by assertion.

"There is massive amounts of evidence for Einstein's theories. There is no evidence for geocentrism."

You are correct. There is no evidence for your libelous term "geocentrism." There is, however, evidence for geoentricity, and your statement is as nonsensical as saying, "Relativity is true, but it is false." If there is no evidence for geocentricity, then Relativity is an invalid theory, because Relativity's geocentric observer must be able to explain observations solely from a geocentric perspective. He could not do that if there were observations that did not support his geocentric perspective.

"Yet there aren't. There are scientists who accept the massive amounts of evidence for Einstein's theories. That doesn't reduce their position to an 'ism.'"

Neither does being a relativist, any more than being a biologist reduces their position to biologism (your word). Why do you hold the ism that any word with "ist" as a suffix automatically reduces to an "ism" when that is demonstrably untrue? Whatever the reason for your relativist-phobia, a person who holds to Einstein's

Relativity is most definitely a relativist.

"Ism' is generally not used for scientific theories. This is because they are not mere opinions. They are models that comprehensive explain observations in reality. Again, fix your lack of knowledge of science. Right now you're just embarrassing yourself."

You are right. I made a poor choice by using 'Geocentrism' rather than the proper and scientific 'Geocentricity.' I admit that, and from now on will only use the term "Geocentricity." Thank you for correcting me.

MomoTheBellyDancer wrote:

"You are correct. Those are all assumptions. Not empirically proven facts. Made by a non-geocentric observer."

I only used the word "assumptions" in a broad sense. We *know* these are facts because they are verified by observations.

"So you're saying you really have no idea how many or how few assumptions the geocentric model allegedly has to make."

I just gave you a list of the assumptions. You're being dishonest again.

""You need to add assumptions fot the seasons," What assumptions are those?"

Geocentric assumption: Some **unknown mechanism** is moving the sun up and down with regard to the earth.

""the retrograde motion of planets," What's the assumption? "

Geocentric assumption: The planets suddenly decide to move backward by circling around some center with an **unknown origin**.

""the precession of the equinox," What's the assumption?"

Geocentric assumption: Somehow, for **reasons unknown**, the stars we observe in the sky slowly shift position in a predictable pattern.

""the phases and transits of Venus and Mercury," No assumptions necessary. You're going with an obsolete geocentric model where Mercury and Venus orbit the Earth rather than the sun."

Geocentric assumption: Some unknown mechanism makes Venus and Mercury rotate arond the sun, along with *other* planets, while the earth is itself fixed. That that whole system then rotates around the earth, even though we **know of no mechanism** by which this could happen. It can't be gravity, since if it were, we'd all be flat as pancakes. Not only that, but some **unknown cause** makes it appear as if the earth is moving with regard to Venus and Mars.

""the existence of the Milky Way," You've actually come up with one I don't think I've ever heard before. I know the one about the sun moving through the Milky Way, but what's the assumption about the existence of the Milky Way?"

You tell me? We see a band of stars across the sky, dispersed with dark bands. According to *your* model, **it's just there**, and it rotates around the earth by a yet completely **unknown mechanism**.

""the Coriolis effect,"What assumption?"

The geocentric assumption is that objects move across the face of the earth with a measurable, predictable variation for **reasons unknown**. Hurricanes in the Northern hemisphere rotate counter-clockwise, and clockwise in the southern hemisphere **for the heck of it**.

"That it's a real force generated by the universe rotating around the Earth?"

No. In reality we know it's a *fictitious* force caused by the rotation of the earth.

"the non-geocentric model makes the assumption that it's a fictitious force."

Which is just there, for completely **unknown reasons**.

""Foucault's pendulum,"Coriolis force from the rotating universe rather than the rotating Earth."

Of course, the universe rotates along the earth by a completely **unknown mechanism**, making far-away stars break the speed of light, for which there is **no explanation** given either. For equally **unknown reasons** pendulums change their rotational speed with altitude as if the earth were a spinning sphere. Right.

"But earthquakes don't shift Earth's axis. They shift the Earth's surface relative to the axis in both the geocentric and non-geocentric view"

So in the geocentric model, localized earthquakes manage to shift the surface of the *whole* planet in one go by the exact same amount, despite the surface being made up of tectonic plates, and without the rest of the planet undergoing massive earthquakes as well. By what mechanism does this happen? **Who knows?** Don't ask questions!

""earth magnetism,"What's the assumption?"

Compass needles align along a magnetic field. Where does this field come from? The geocentric explanation is : *durrr*. **It's just there**. *Don't ask questions!*

All the parts I put in bold are the extra variables you have to introduce into the system in order to make the geocentric model work. These are all readily explained on a spherical moving earth, whereas geocentrists only respond with a massive shrug.

"Astronomical observation shows that everything is generally receding from Earth (red shift). This implies we're at the center of the universe."

Wow, you're stupid. No, we're not at the center of the universe. The universe is equally receding from *every* point in the universe. This is because the universe is a four-dimensional construct, and spacetime *itself* is expanding.

"This hypothesis is only true if an observer at the edge of Earth's observable universe (our horizon, as you probably call it) can see something beyond the edge (or horizon)."

Have you ever heard of this thing called a*telescope*? It enables to peek way into the universe past this so-called horizon of yours.

"If there's nothing beyond the edge (or horizon"

But there is. We can see it.

"And if we're literally at the center of the universe"

But we're not.

"then Earth is in a preferred, unique position, establishing an absolute reference frame"
It isn't.

"rendering Relativity invalid."

Please contact CERN and explain them that the equations they use during their experiments don't work. Be prepared to be laughed out of court.

Heck, tell the companies that make GPS receivers to stop compensating for relativistic effects in their software. See how *that* goes.

Scott Reeves wrote:

"We *know* these are facts because they are verified by observations."

I quote your god Einstein: "The second class of facts to which we have alluded has reference to the question whether or not the motion of the Earth in space can be made perceptible in terrestrial experiments. We have already remarked in section 5 that all attempts of this nature led to a negative result." (Albert Einstein, Relativity, section 16).

What does he say in section 5? "If the principle of relativity were not valid we should therefore expect that the direction of motion of the earth at any moment would enter into the laws of nature, and also that physical systems in their behavior would be dependent on the orientation in space with respect to the earth...However, the most careful observations have never revealed such anisotropic properties in terrestrial physical space." (ibid.)

Simply put, what you claim as facts are only facts from the viewpoint of a non-geocentric observer. If you can say definitively that Earth is orbiting the sun, then, as Einstein says, the principle of relativity is not valid, and (as I say) his theory belongs in the dustbin of history.

"I just gave you a list of the assumptions. You're being dishonest again."

No, I have never said anything dishonest in our back and forth. You claimed there were anywhere from a dozen to hundreds of assumptions. You gave me a list of nine assumptions. Are you now claiming that it is your complete list? Because if it is, you lose, since none of the "assumptions" on your list are strikes against geocentricity.

"Geocentric assumption: Some **unknown mechanism** is moving the sun up and down with regard to the earth."

It's not an unknown mechanism. It's caused by way mass is distributed in the universe.

"Geocentric assumption: The planets suddenly decide to move backward by circling around some center with an **unknown origin**."

No. Only in an obsolete geocentric model that has the planets orbiting the Earth rather than the sun. Besides, the planets only appear to move backward on the celestial sphere. Which they do in the heliocentric model as well. They appear to move backward on the celestial sphere no matter what sort of model of the universe you use. Which is why any model of the universe must account for the motions as observed on the celestial sphere.

"Geocentric assumption: Some unknown mechanism makes Venus and Mercury rotate arond the sun,"

It's not an unknown mechanism. It's called gravity.

"along with *other* planets, while the earth is itself fixed. That that whole system then rotates around the earth, even though we **know of no mechanism** by which this could happen."

We do know of such a mechanism. If you have a system of bodies, that system of bodies revolves around their common center of mass. The whole universe is a system of bodies. In the heliocentric model, the sun itself only revolves around the solar system's barycenter if you consider the solar system to be isolated from the rest of the universe. Otherwise the sun revolves around some barycenter external to the solar system. When you consider the visible universe as a whole, that external barycenter is centered on Earth.

"It can't be gravity, since if it were, we'd all be flat as pancakes."

We've already had that argument in the comment thread of some other video. Proof by assertion.

"Not only that, but some **unknown cause** makes it appear as if the earth is moving with regard to Venus and Mars."

That's because the earth does appear to be moving with regard to Venus and Mars. Or Venus and Mars appear to be moving with regard to earth. It's called relative motion. But the cause is not unknown. It's called gravity.

"We see a band of stars across the sky, dispersed with dark bands. According to *your* model, **it's just there**, and it rotates around the earth by a yet completely **unknown mechanism**."

Barycenter of the universe again. You're just repeating an earlier alleged assumption, using the Milky Way this time rather than the solar system. As for whether it's "just there," I said earlier that I'm not making any claims about the origin of the universe or anything within it. Your model of the universe doesn't have any definitive and non-speculative answer for how our universe is here either.

"The geocentric assumption is that objects move across the face of the earth with a measurable, predictable variation for **reasons unknown**. Hurricanes in the Northern hemisphere rotate counter-clockwise, and clockwise in the southern hemisphere **for the heck of it**."

It's not an unknown reason. It's caused by the rotation of distant masses around the Earth.

"No. In reality we know it's a *fictitious* force caused by the rotation of the earth."

It's a fictitious force from the viewpoint of a non-geocentric observer. It's a real force caused by the rotation of the distant stars from the viewpoint of a geocentric observer. Why do you as a geocentric observer keep insisting upon assuming the viewpoint of a non-geocentric observer?

In his "Dialogue About Objections Against the Theory of Relativity" (where, FYI, there are multiple references to your allegedly non-existent relativist), Einstein argues against the classification of forces as being real or fictitious, because he knows that such a classification destroys his theory. If you are an adherent of Relativity, then you should resist that classification as well. By making such a classification, you are actually arguing against Relativity.

"Which is just there, for completely **unknown reasons**." [In response to *"the non-geocentric model makes the assumption that it's a fictitious force."*]

It's not there for completely unknown reasons. According to the non-geocentric observer, it's there due to the rotation of the Earth. Why do you believe that a geocentric observer can't explain the Coriolis Effect from a non-geocentric viewpoint, and vice-versa?

"Of course, the universe rotates along the earth by a completely **unknown mechanism**, making far-away stars break the speed of light, for which there is **no explanation** given either. For equally **unknown reasons** pendulums change their rotational speed with altitude as if the earth were a spinning sphere. Right."

Throughout your replies, you keep referring to unknown mechanisms, lack of explanations, and unknown reasons. There is nothing unknown about the mechanisms and reasons, and the explanations are given if you truly care to dig deeply into the subject of geocentricity.

"So in the geocentric model, localized earthquakes manage to shift the surface of the *whole* planet in one go by the exact same amount, despite the surface being made up of tectonic plates, and without the rest of the planet undergoing massive earthquakes as well. By what mechanism does this happen? **Who knows?** Don't ask questions!"

This isn't an explanation concocted by absolute Geocentrists. This explanation comes from mainstream scientists. Don't shoot the messenger. The short answer is that earthquakes affect the axial tilt, but the long answer is that earthquakes only affect the location of the axis relative to observers on the surface of the Earth due to shifting of the crust. The tilt relative to the stars and planets of the universe is unchanged. Do a bit of research.

[LATER NOTE: This was actually a poor explanation of it on my part. It is true that the axis of rotation doesn't change relative to the distant stars, but rather than the crust shifting relative to the axis of rotation isn't entirely accurate. What IS accurate is that earthquakes change the distribution of Earth's mass around its figure axis, which is not the same as the rotational axis. Which still changes how the Earth wobbles when it rotates, or alternatively, how the universe wobbles when it rotates around Earth. But earthquakes DO NOT affect Earth's axis of rotation, or alternatively, the universe's axis of rotation.

NASA says this about the subject at

http://www.nasa.gov/topics/earth/features/japanquake/earth2011 0314.html:

"The calculations also show the Japan quake should have shifted the position of Earth's figure axis (the axis about which Earth's mass is balanced) by about 17 centimeters (6.5 inches), towards 133 degrees east longitude. Earth's figure axis should not be confused with its north-south axis; they are offset by about 10 meters (about 33 feet). This shift in Earth's figure axis will cause Earth to wobble a bit differently as it rotates, but it will not cause a shift of Earth's axis in space—only external forces such as the gravitational attraction of the sun, moon and planets can do that."

The above link has been saved to the Internet Archive's Wayback Machine in case it is no longer working or has been modified after publication of this book (yes, of course I am paranoid about that happening). More from the same article:

"Gross said that while we can measure the effects of the atmosphere and ocean on Earth's rotation, the effects of earthquakes, at least up until now, have been too small to measure. The computed change in the length of day caused by earthquakes is much smaller than the accuracy with which scientists can currently measure changes in the length of the day. However, since the position of the figure axis can be measured to an accuracy of about 5 centimeters (2 inches), the estimated 17-centimeter shift in the figure axis from the Japan quake may actually be large enough to observe if scientists can adequately remove the larger effects of the atmosphere and ocean from the Earth rotation measurements. He and other scientists will be investigating this as more data become available."

I don't know about other readers, but what the preceding paragraphs, straight from NASA's own website, say to me is that not only do earthquakes not affect the rotational axis, it's never actually been proven that they even affect the figure axis. It's only been CALCULATED that the distribution of mass around the figure axis has changed.

Absolute Geocentrists have much to say on the subject of earthquakes and the rest of the universe, a subject that is too detailed and involved to go into here. But if any readers are truly interested, the subject of earthquakes is not the strike against geocentricity that its opponents would have you believe, and the resources are out there to learn more about the geocentric position on earthquakes.

Here is another response I had ready in case the earthquake argument was pursued further:

Fine. The only explanation from a geocentric perspective is that earthquakes instantly affect the entire universe. At this point, you have two options: 1) you can say that invalidates a geocentric universe, which also invalidates Relativity, since your implicit claim is that the rotation of the Earth is the best explanation for earthquakes, which is an appeal to the viewpoint of a non-geocentric observer for an explanation, which means that the geocentric reference frame is an inferior reference frame, which means that it is not equivalent to all other frames, which violates Relativity, and that violation invalidates Relativity. So option 1) means that earthquakes are a disproof of Relativity.

Option 2) is to be a true adherent of Relativity and say, "Wow! Earth is able to instantly influence the entire universe! That's amazing!" And then as a scientist you cream in your jeans because a whole new field of scientific research has been opened up.

Either way, I win, because if you take the first option, my prime contention all along has been that Relativity is an invalid and ridiculous theory, and you're finally validating my contention. If you take the second option, you still haven't demonstrated that we don't live in an absolute Geocentric universe, so your argument against me remains unchanged. Earthquakes are a win-win situation for me, so what do I care about your earthquake argument? If you want to remain a Relativity adherent, and you don't like the notion that earthquakes instantly affect the entire universe from a geocentric perspective, then get out there and find an alternative explanation you're more comfortable with. I'm not the one who thinks earthquakes are a problem for the geocentric model.]

"Compass needles align along a magnetic field. Where does this field come from? The geocentric explanation is : *durrr*. **It's just there**. *Don't ask questions!*"

Sounds to me like this is just another manifestation of the Coriolis Effect, which in the geocentric model is caused by the rotation of distant masses around the Earth.

"All the parts I put in bold are the extra variables you have to introduce into the system in order to make the geocentric model work. These are all readily explained on a spherical moving earth, whereas geocentrists only respond with a massive shrug."

Then you don't have to introduce extra variables. Because pretty much everything you presented is explicable in terms of the Coriolis Effect, and the true question is whether it's a real or fictitious force, and the correct answer, entirely consistent with both absolute Geocentrists and relativists is, "It depends upon which observer you ask," or, as Einstein says, "the distinction real-unreal is hardly helpful." (DAOATTR). You'll note I didn't respond with a massive shrug, and if I missed some objection you were raising, the answers are out there if you make an effort to seek them out.

"Wow, you're stupid."

Incorrect. You are having an encounter with a bona fide genius. You don't appreciate how fortunate you truly are.

"No, we're not at the center of the universe."

Mainstream scientists most definitely do concede that Earth is at the center of its observable universe.

"The universe is equally receding from *every* point in the universe."

See? Not only do mainstream scientists concede it, but you do as well.

Anyway, that's an interesting hypothesis. Now go out to a statistically significant number of points in the universe and test it. Additionally, one of the implications of your statement is that Earth is at the center of its own observable universe, and that there are no privileged observers (Relativity). In order for your statement to be true, you must retract all your objections against the geocentric model and agree that everything I've been saying about both Relativity and geocentricity is correct. Are you prepared to do that? Repeat after me: "We are most definitely in a geocentric universe. The only debate is whether it is relativistically geocentric, or absolutely Geocentric."

Only when someone repeats that statement without the slightest bit of hesitation or disagreement with it can he/she be considered to truly understand Relativity. You can and will, I'm sure, insist upon adding the caveat that there is no debate, and I will then add the same caveat, and we will continue debating. We will just focus the debate entirely upon the validity of Relativity, because that has been the real debate all along. If you truly understood Relativity, you would have been fighting tooth-and-nail in favor of the geocentric model all along, and insisting that we instead shift the discussion to Relativity. Which means that we would have shifted to Relativity at the very beginning of our discussion, since I knew from the outset that that was the real debate. All along, I've merely been trying to get you to see where the true debate lies.

"This is because the universe is a four-dimensional construct, and spacetime *itself* is expanding."

Another interesting hypothesis. I've also heard that we're in a ten-dimensional universe. There are all sorts of ideas about what sort of universe we live in and how many dimensions it has. And the absolute Geocentric one has the most empirical support.

"Have you ever heard of this thing called a *telescope*? It enables to peek way into the universe past this so-called horizon of yours."

No it doesn't. There's a reason it's called a horizon. You know, that thing you can't see beyond? And it's not my horizon, since I hypothesize that it's an edge, not a horizon. It's Relativity's horizon. And since according to mainstream scientists, the universe is 13.8 billion years old, we can't see anything beyond that distance. It's pure speculation that anything exists beyond that distance, despite the untested hypothesis that the observable universe is 46 billion light years in radius. How do you see beyond a typical horizon? You move to a vantage point closer to the horizon. Observers on Earth have not yet done that. Or you query observers that have been over the horizon. Observers on Earth have not yet done that.

"But there is. We can see it."

You cannot possibly be serious.

"But we're not."

We are if Relativity is false or if there is nothing beyond 13.8 billion light years. And you have no evidence that there is.

"It isn't."

It is if Relativity is false or there is nothing beyond 13.8 billion light years. And you have no evidence that there is.

"Please contact CERN and explain them that the equations they use during their experiments don't work. Be prepared to be laughed out of court."

Which equation? The time dilation equation? I don't preclude the possibility that time dilation exists (not entirely, but that would be too lengthy to go into right now), only that, if it exists, it is not reciprocal as Relativity hypothesizes. So I have no problem with accelerated particles undergoing time dilation. Particle accelerators prove that strange things happen when you move relative to the Earth, which is what you would expect if Earth is a privileged reference frame. Now set up particle accelerators on other planets and in deep space and demonstrate that you get the results you expect without taking into account the motion of the particles relative to Earth.

"Heck, tell the companies that make GPS receivers to stop compensating for relativistic effects in their software. See how *that* goes."

Why would I do such a thing? Those corrections are necessary to the system because they're not relativistic effects, since absolute Geocentricity predicts the need for them as well. You know that so-called time dilation equation? There's nothing inherently relativistic about it. Take a look at it and point out precisely where it is relativistic. I predict that you're going to point at c and tell me that c is constant for all observers, but that's not what the equation says. That's how Einstein says we're supposed to interpret the equation when he incorporates what he views as time dilation into Relativity. I can derive the exact same time dilation equation using the exact same logic and method as Relativity. What makes time dilation relativistic is that Einstein claims it is symmetric, which absolute Geocentrism denies. To claim GPS as support for Relativity, go set up a GPS system on Mars or elsewhere and show that you don't have to take the motion of the atomic clocks relative to the Earth into account. As for the gravitational component of the corrections – nothing inherently relativistic about that correction either.

MomoTheBellyDancer wrote:

"Mainstream scientists most definitely do concede that Earth is at the center of its observable universe"

Citation needed.

"It's Relativity's horizon"

Citation needed.

"It's pure speculation that anything exists beyond that distance, "

Citation needed.

"Particle accelerators prove that strange things happen when you move relative to the Earth"

Citation needed.

"Those corrections are necessary to the system because they're not relativistic effects,"

Citation needed.

"Take a look at it and point out precisely where it is relativistic"

http://www.astronomy.ohio-state.edu/~pogge/Ast162/Unit5/gps.html

"As for the gravitational component of the corrections – nothing inherently relativistic about that correction either."

Citation needed.

"Relative motion is not a concept that is exclusive to Einstein's theory, nor is what he calls time dilation."

Citation needed.

"The GPS system would still require corrections based upon the same equations if there were an absolute reference frame."

Citation needed.

Scott Reeves wrote:

"Citation needed [for 'Mainstream scientists most definitely do concede that Earth is at the center of its observable universe']"

It's common knowledge, so common that citation is only needed for science dolts or people who are being contentious just to be contentious. You yourself implicitly admitted that you agree with my statement when you said, "The universe is equally receding from *every* point in the universe." That's a paraphrase of the Copernican Principle. The observational evidence shows that Earth is the center of the universe, which is why scientists interpret the evidence through the filter of the Copernican Principle, and thus say, as you did, that "The universe is equally receding from *every* point in the universe." Which is an interesting hypothesis. Now go test it.

"Citation needed [for 'It's Relativity's horizon.']"

No citation needed. I said that it's not my horizon. If I say that something doesn't belong to me, I don't need to cite an external source. As far as it being Relativity's horizon, if Relativity doesn't claim that our observable universe has a horizon, then Relativity must claim our observable universe has an edge or boundary, and if our universe has such an edge or boundary, then being at the center of the observable universe as we observably are, that means we are literally at the center of the entire universe and Relativity is false. Therefore what I call an edge, Relativity must necessarily call a horizon. If you aren't capable of seeing the logical truth of this on your own without having to be pointed to an external source for corroboration, I don't know what to tell you.

"Citation needed [to 'It's pure speculation that anything exists beyond that distance']"

This is like saying that citation is needed for a claim that fairies and unicorns do not exist. Saying that anything definitely exists in a region that cannot be empirically observed can't possibly be anything BUT speculation.

"Citation needed [for 'Particle accelerators prove that strange things happen when you move relative to the Earth']"

Citation not needed. Particle accelerators do prove that strange things happen when you move relative to the Earth. Unless you don't think the effect that Einstein calls time dilation "strange." Maybe you would use the word "interesting" instead. Whatever you call it, particle accelerators prove that something weird/strange/interesting/[insert your adjective of choice] happens to particles in motion relative to the Earth. This cannot be denied. What CAN be denied is that that something is explicable solely by Relativity, which is what I am doing, and the fact that I'm denying something needs no citation.

"Citation needed [for "Those corrections are necessary to the system because they're not relativistic effects]"

Okay, I'll grant you that this one would need citation, because I realize in retrospect that I misstated what I actually meant. I shouldn't have said they weren't relativistic effects, because they clearly are – but only according to Relativity. What I actually meant was that the effect (the effect that Relativity labels as time dilation, whether due to relative motion or gravity) is not something that is explicable solely by Relativity.

"http://www.astronomy.ohio-state.edu/~pogge/Ast162/Unit5/gps.html [for 'Take a look at it and point out precisely where it is relativistic']"

Same answer as above. I misspoke regarding time dilation not being a relativistic effect. It is relativistic – when you explain it according to Relativity's interpretation.

But: nothing on the page you've cited refutes what I said about the need for GPS corrections not being exclusive to Relativity. It merely confirms that the necessity for GPS corrections is predicted by Relativity, which is not in dispute. And if I gave the impression that such was in dispute, it's because I mis-stated my position as above. The real dispute is over the claim that nothing outside Relativity can predict the need for the exact same corrections, using the exact same equations as Relativity, and the cited article does not in any way settle or even address that dispute. It merely explains the GPS system from the perspective of Relativity.

"Citation needed [for 'Relative motion is not a concept that is exclusive to Einstein's theory, nor is what he calls time dilation.']"

Regarding relative motion, citation not needed, unless you've never heard of Galilean relativity, in which case you're most definitely not qualified to be debating this topic. But I doubt that you haven't heard of it. As for my claims about time dilation, pretend that I'm the first to ever try to wrest the concept of time dilation away from Einstein, so there are no outside sources to cite, and I have not yet submitted a paper for peer-review, and I'm merely looking for you to informally point out where I'm going wrong. Look at the time dilation equations for both types of time dilation. Look at each one carefully and point out where a uniquely Einsteinian variable or concept is embodied within the math, without resorting to an external verbal, written or mathematical stipulation that the speed of light is constant for all observers. I am utterly and supremely confident that you cannot do it. Demonstrate how my claim is incorrect.

"Citation needed [for 'The GPS system would still require corrections based upon the same equations if there were an absolute reference frame.']"

No citation needed. It's common knowledge among scientifically literate people that the so-called time dilation equation is actually the Lorentz factor, which predates Relativity by a good decade, and was originally proposed to explain the inability of Michelson-Morley to detect motion relative to an absolute reference frame (the ether). No citation needed, this is a matter of basic science history that you surely know since you are scientifically literate. Einstein incorporated the equation, unchanged, into Relativity, and simply declared that the velocity of light was constant for all observers rather than constant relative to an absolute reference frame. Since the equation is unchanged from its original formulation, and the original formulation assumed an absolute reference frame, it is still applies to an absolute reference frame, which is what absolute Geocentrists consider the Earth's frame to be. Hence, using the exact same equation as Relativity, absolute Geocentricity predicts the needed correction to the GPS system. As for the need for correcting for gravitational time dilation, once you predict what Relativity calls time dilation, it's just a short mental leap to predict gravitational time dilation. Since two clocks are initially in synch, and motion gets them out of synch, and you realize that gravity causes things to accelerate (which was realized long before Einstein, of course), you realize that someone closer to a large mass is experiencing a greater acceleration than someone more distant, and thus you would conclude that the closer observer is experiencing so-called time dilation to a greater degree than a more distant observer. Thus absolute Geocentricity also predicts that an atomic clock high above the Earth would be the equivalent of a faster-ticking clock, and corrections would be needed to the GPS system using a gravitationally-based equation.

MomoTheBellyDancer wrote:

"It's common knowledge,"

Useless.

"The observational evidence shows that Earth is the center of the universe"

Citation needed.

"if Relativity doesn't claim that our observable universe has a horizon then Relativity must claim our observable universe has an edge or boundary"

Non sequitur.

Also, citation needed.

"Saying that anything definitely exists in a region that cannot be empirically observed can't possibly be anything BUT speculation."

Citation *still* needed.

"Particle accelerators do prove that strange things happen when you move relative to the Earth."

Citation *still* needed

"the effect (the effect that Relativity labels as time dilation, whether due to relative motion or gravity) is not something that is explicable solely by Relativity."

Citation needed.

"there are no outside sources to cite"

Which renders your claim worthless.

"Einstein incorporated the equation, unchanged, into Relativity, and simply declared that the velocity of light was constant for all observers rather than constant relative to an absolute reference frame"

Which is an extremely important difference, which is backed up by massive amounts of evidence.

"absolute Geocentricity also predicts that an atomic clock high above the Earth would be the equivalent of a faster-ticking clock"

<u>Scott Reeves wrote:</u>

"Citation needed [for 'The observational evidence shows that Earth is the center of the universe']"

https://www.google.com/search?q=earth+is+the+center+of+the+observable+universe

https://www.google.com/search?biw=1280&bih=685&q=hubble%27s+law&oq=hubble%27s+law

https://en.wikipedia.org/wiki/Hubble_volume

Your sudden constant need for citations is merely your way of avoiding the fact that you are unable to refute my arguments. The observational evidence does show that Earth is the center of the universe. You yourself implicitly acknowledged this when you earlier said, "The universe is equally receding from *every* point in the universe." Your statement would be incorrect if the observational evidence did not show that Earth is the center of the universe. Your statement implies that any observer in the universe will have observational evidence showing that his/her/its location is the center of the universe, including an observer on Earth. Your statement is in fact a hypothesis put forth in an attempt to interpret the observational evidence in a non-geocentric fashion. One of the implications of your hypothesis is that a larger universe exists beyond the universe that we can observe from Earth. Again, go ye forth and test your hypothesis. Travel to the edge of Earth's observable universe, set up one of them telescope thingamajigs you mentioned earlier, and see if you can observe anything beyond the edge of Earth's observable universe. Or wait until Earth is visited by beings from a planet on edge of Earth's observable universe and ask them to provide us with a record of astronomical observations from their homeworld, examine those records, and see if they show a universe beyond the edge of Earth's observable universe.

"Non sequitur. [for 'if Relativity doesn't claim that our observable universe has a horizon then Relativity must claim our observable universe has an edge or boundary']"

If it's a non sequitur, then YOU finish the sentence in a logical way: 'If Relativity doesn't claim that our observable universe has a horizon then Relativity must claim...?' Our observable universe MUST have either a horizon or an edge. If you believe there's a logical alternative to an edge or a horizon, then insert that alternative into the above unfinished sentence.

"Also, citation needed [for 'if Relativity doesn't claim that our observable universe has a horizon then Relativity must claim our observable universe has an edge or boundary.']"

Citation not needed, argumentum ad avoidingum issuem. It's completely obvious to anyone with at least half a brain. If you disagree with this statement: "If Earth's observable universe is the entire universe, and if Earth is at the center of its observable universe, then Relativity is an invalid theory," you are an idiot. You have already implicitly agreed that Earth is at the center of its observable universe when you said, "The universe is equally receding from *every* point in the universe." Thus there are only two points left in the above statement on which to disagree: 1) Earth's observable universe is the entire universe, and 2) whether Earth being at the center of the entire universe would mean that Relativity is an invalid theory. If you disagree that Earth being at the center of the entire universe would invalidate Relativity, then you don't understand Relativity. That leaves us only one point upon which to disagree: that Earth's observable universe is the entire universe. If the edge of Earth's observable universe is an actual edge, meaning the universe ends there, then Earth is literally at the center of the universe. Thus, for Relativity to be a valid theory, the edge of Earth's observable universe cannot be an actual edge. Thus, it must be a horizon, meaning there's more to the universe, we just can't see it from Earth.

"Citation still needed [for 'Saying that anything definitely exists in a region that cannot be empirically observed can't possibly be anything BUT speculation.']"

You want me to cite evidence to back up a claim that if something is empirically unobservable, then that something's existence is purely a matter of speculation? No. I'm not going to cite evidence for a statement that only someone who is completely ignorant of the scientific method would claim is untrue.

"Citation needed [for 'the effect (the effect that Relativity labels as time dilation, whether due to relative motion or gravity) is not something that is explicable solely by Relativity.']"

Citation not needed, unless you are unable to examine the Lorentz factor (Einstein's time dilation equation) or the gravitational time dilation equation on your own. My challenge stands. Examine the so-called time dilation equation (aka the Lorentz factor) at least and point out, based solely upon the math of the equation itself, without resorting to the extra-equational stipulation that the velocity of light is constant for all observers, exactly where the equation has no other interpretation but an Einsteinian one.

The mere fact that the equation predates Einstein by at least a decade, and was being interpreted with the assumption of an absolute reference frame (the ether) proves that it has a non-Einsteinian interpretation. But if you continue your ridiculous insistence upon citation for things which you as an allegedly scientifically literate person should already know, here is one citation: https://en.wikipedia.org/wiki/Lorentz_ether_theory

If Wikipedia is not an acceptable source for you, too bad. If you truly need citation for such basic matters, then dig up others on your own time. There are any number of other sources out there, as what I'm saying is a matter of accepted historical fact.

"Which renders your claim worthless. [to 'there are no outside sources to cite']"

I see you're wearing your OSHA-approved quote-miner's helmet. The full statement was: "As for my claims about time dilation, pretend that I'm the first to ever try and succeed at wresting the concept of time dilation away from Einstein, so there are no outside sources to cite..." I didn't say there were no sources to cite. I asked you to pretend that there weren't and refute my claim regarding the "time dilation" equation using your own knowledge and intelligence, which you appear unwilling or unable to do. I'm going with unable, because it is irrefutable that the "time dilation" equation is not inherently relativistic.

"Which is an extremely important difference, which is backed up by massive amounts of evidence."

Yes, it is an extremely important difference, and it's the difference between a universe with an absolute reference frame and a universe with no absolute reference frame. But that difference makes no difference to my claim that the "time dilation" equation is not the sole property of Relativity, and my challenge still stands. Show me, based solely upon the internal math of the equation, what makes this equation a solely Einsteinian equation. Based upon the math of the equation by itself, you CANNOT show that it is only explicable in terms of Relativity, which is why you are avoiding the challenge.

As for the massive amounts of evidence: what sort of evidence? Evidence such as the fact that particle accelerators demonstrate that strange things seem to happen when you move relative to the Earth? Why do you insist upon a citation for a claim that as a relativist you should agree with? Unless you merely take issue with the description of "time dilation" being described as "strange." And anyway, the truth of your statement "backed up by massive amounts of evidence" is an inextricable part of the debate we're having. Your "massive amounts of evidence" also support geocentricity. They must, or Relativity is false. We indisputably live in a geocentric universe. You have yet to understand that the only disagreement is whether we live in a relativistically geocentric universe, or an absolutely Geocentric universe.

[I am aware of no further response as of date of publication of this book]

Debate Eight

Scott Reeves vs. Earthbound

Earthbound wrote:

Scott: Do you really think that the Bible and the Coran were intended to prove that people who are capeable of proving just how "incredibly" wonderous the Universe really is are all "idiots"? I'd love to see you Google a physics course by, let's say? Susskind at Stanford and then explain to us "in plain language" what you've understood? Opinions are a dime a dozen. Wisdom and knowledge are things that take an effort to comprehend.

Scott Reeves wrote:

"Do you really think that the Bible and the Coran were intended to prove that people who are capeable of proving just how "incredibly" wonderous the Universe really is are all "idiots"?"

No, I don't really think that, and I don't recall ever saying that I did think that. Unlike most geocentrists, I don't use God or the Bible in my arguments for absolute Geocentrism.

"I'd love to see you Google a physics course by, let's say? Susskind at Stanford and then explain to us "in plain language" what you've understood?"

I Googled it, and I understand that he has a boatload of physics lectures available on YouTube. And I understood this even before I Googled him.

But I assume you meant you'd love to see me watch his physics lectures and then summarize them in my own words. I will not be doing that, since summarizing 100 lectures each running about an hour and a half would be a more time-consuming task than I'm willing to undertake. Not opposed to watching them, just to summarizing them just in the hopes that you or others might possibly find something in the summaries to use as a demonstration that I didn't understand what I watched. Besides, I've already watched numerous physics lectures on YouTube in the past, possibly even Susskind's, though I don't recall watching his particular lectures. I've read countless books on relativity and quantum mechanics, watched tons of videos, taken physics and other science classes at university, etc. And all of that has led me to absolute Geocentrism. Religion didn't lead me here, decades of studying science did.

"Opinions are a dime a dozen."

Yes they are. But the majority of my comments and arguments don't involve opinions. They involve advocating a different, non-mainstream interpretation of the same scientific facts and observations to which everyone has access.

"Wisdom and knowledge are things that take an effort to comprehend."

Which is why I've spent more than twenty years now, and have sacrificed a ridiculous and obsessive amount of each day, making such an effort. And the effort has paid off.

Earthbound wrote:

Could you explain how GPS systems work without atomic clock corrections -- or do you also deny their existence too (lol) Before Arthur C. Clark collaborated on the script for Kubrick's 2001 Space Odysee, they spent six years interviewing NASA experts -- you know those gentlemen Russ says have spent all these years since 1957 and the Russian Sputnick -- probably also a Russian conspiracy -- making "fake" movies so Russ can "denounce" them and make money, himself, off of YouTube clicks. Russ may not know much about science, but he does apparently understand how Google makes a living. And thank you, at least, for giving a good laugh to my theoretical physicist companion. He never ceases to be amazed at how naive we Americans can be?

Scott Reeves wrote:

"Could you explain how GPS systems work without atomic clock corrections -- or do you also deny their existence too (lol)"

No, I could not explain how GPS systems work without atomic clock corrections, and no, I don't deny their existence. The need for GPS atomic clock corrections is predicted by absolute Geocentrism as well, and they're predicted according to the same reasoning and equation that Einstein used. Look at the so-called time dilation equation and show me what makes it exclusive to Relativity? What Einstein labeled time dilation is not an inherently relativistic concept. It only becomes a relativistic concept when Einstein hypothesizes that time dilation is symmetric. To properly claim GPS as a test of Relativity, go set up a GPS system on Mars or some other planet. The absolute Geocentric model (according to this absolute Geocentrist at least) predicts that for GPS to work on Mars or elsewhere, the velocity of the satellites relative to the absolute reference frame (Earth's reference frame in the absolute Geocentric model) will have to be taken into account, rather than its velocity relative to an atomic clock stationary on the surface of Mars, as relativity predicts.

"Before Arthur C. Clark collaborated on the script for Kubrick's 2001 Space Odysee, they spent six years interviewing NASA experts -- you know those gentlemen Russ says have spent all these years since 1957 and the Russian Sputnick -- probably also a Russian conspiracy -- making "fake" movies so Russ can "denounce" them and make money, himself, off of YouTube clicks."

The presence of my comments on Russ's video page does not constitute my agreement with Russ's ideas any more than the presence of your own comments on the same video page does, so I'm not going to defend his position regarding Russians making fake movies. BUT – consulting with NASA for six years to make a science fiction movie isn't a valid test of any scientific theory, as far as I'm aware. And 2010 was a much better movie than 2001.

"And thank you, at least, for giving a good laugh to my theoretical physicist companion. He never ceases to be amazed at how naive we Americans can be?"

You're welcome. If nothing else, it's nice to know that I can at least give the gift of laughter.

Indra wrote:

+Earthbound

As far as I know, GPS system requires atomic clock correction but these corrections has nothing to do with special and general theory of relativity. In fact GPS satellite data contradicts Einstein's theory of relativity. Ronald R. Hatch who is the Director of Navigation Systems at NavCom Technology and a former president of the Institute of Navigation has stated this clearly in his book "Escape from Einstein". It is amazing how much myth is perpetuated by popular science gurus and dumb physicists who repeat things without ever investigating them. [Escape-Einstein-Ronald-R-Hatch – linked removed to make eBook version of this book retailer friendly]

Earthbound wrote:

+Indra

Sorry, Indra, I just happen to live with one of those "dumb" people called physicists and have direct links to two nobel prize winners at UC Berkeley. Maybe you are up on what popular science gurus and others here are wasting our time perpetuating. Frankly, after seeing Russ's other video, I'm calling it a day and will let you go on debating with the "well-informed" people who are commenting here. I'd rather listen to those who really are "well" informed enough to laugh at yours and the others nonsense.

Scott Reeves wrote:

"Frankly, after seeing Russ's other video, I'm calling it a day and will let you go on debating with the "well-informed" people who are commenting here. I'd rather listen to those who really are "well" informed enough to laugh at yours and the others nonsense."

And that's part of the problem of modern science. Most people don't want to hear ideas that are at odds with the worldview of the majority of scientists. Let's only listen to people who adhere to the standard theories, and laugh at anyone who dares to point out that there are valid alternative interpretations of the evidence. Anyone who doesn't adhere to the standard majority dogma is obviously an idiot and doesn't understand the accepted theories, because if they truly understood the accepted theories, then they wouldn't reject them.

Indra wrote:

+Earthbound

I have nothing against scientists who are working hard at LIGO or other similar organization. They are working hard for what they believe in and I respect that. These scientists are sometime nothing but pawns in the hands of powerful scientific establishments that hold onto their dogma religiously. I think you are well informed and I seriously doubt you live with a dumb physicist :). I did not mean to offend anyone - just wanted to point out the fact that many things told by popular science gurus on the Internet are not true. And many established theories in physics needs major overhaul or abandonment because they are failing miserably in the face of real experimental data. I am sure the smart physicist you live with will agree with me :).

Scott Reeves wrote:

+Indra

I agree with everything you said, except for "I am sure the smart physicist you live with will agree with me :)." I predict that he would not agree.

Earthbound wrote:

+Indra

I've vowed to stop this useless polemic, but am stuck in a habit I've acquired over the many years observing the US from abroad -- I plunged back in from 1993 to 2002 and faced with the hysteria everywhere following 9/11 interrupted by sabbatical and came rushing back to my teaching post here in #### -- of wondering at how a country with so many sources of information and educational opportunities can remain so profoundly "ignorant"? Maybe it looks "too easy" to too many (or too hard) to make the effort? It's true that I found that most of my US students considered that they could always "look up the information" (and often copy without understanding, just to get a paper in) -- the big difference between having a lot of books on a bookshelf and actual knowledge. Having taught to many engineers, here in ####, who work regularly with their US colleagues, none can understand why so many American engineers are trained to carry out programs and are totally unable to problem solve -- the US company has to send their personnel to #### to find solutions! Our son (born in ####) accompanied me back to my hometown in #### and has stayed there half of his life now, long enough to set up three media and communication companies. I don't doubt the flexibility that is the basis of the "American Dream", just find tragically sad that "opinion" is so often confused with "knowledge" or even with interest in understanding what is really going on, be it in whatever field, for example history (other than US) or geography (other than US), not just in science? Even the BBC news is centered on the US and, at that, how many students know what year the Civil war began or ended (or even care)?

Earthbound wrote (in a different comment thread):

When you're able to explain the formulas envolved and then debunk them, we'll stop laughing. As for the religious tenant of geocentrism, fanatics finally stopped burning those who "dared" give proof to the contrary centuries ago and DAECH is now back cutting throats to "prove" the same theories,right? Are you sure you know whose side you're on?

Scott Reeves wrote:

Who was burned for arguing against geocentrism hundreds of years ago? As for DAESH "cutting throats to prove the same theories," I don't think those scumbags are cutting throats to prove geocentrism. I certainly wouldn't claim such a barbaric act constitutes proof of geocentrism. And I also wouldn't claim that their allegedly doing so constitutes a disproof of geocentrism, any more than Hitler's belief in Relativity would constitute a disproof of Relativity.

Earthbound wrote:

It's great to be a Star Trek fan, to love writing self-published books and, as Michio Kaku puts it, maybe one of these days when (and if) we survivre long enough to move out of a "Zero" level of energy evolution to a third level (Star Trek) stage and understand more about what really takes place in those black holes -- that's what "serious" research is all about? Meanwhile, may I suggest, for example, Sleepwalkers by Arthur Koestler. It's true that he was also fascinated by sci-fi and also by Carl Jung, but mostly the book is interesting in so far as it gives one a little insight concerning the history of cosmology?

Scott Reeves wrote:

Serious research is also "all about" being open to the possibility that by the time you reach that third level (Star Trek) stage, serious research might have revealed that black holes do not exist and that we live in an absolutely Geocentric universe. Serious research doesn't assume it's an indisputable fact that black holes exist and that we don't live in an absolutely Geocentric universe, and then inflexibly hold to that "fact" and interpret any evidence based upon the truth of it. That sort of inflexibility is what leads to ad hoc, save-the-theory-by-pulling-something-out-of-your-butt things like dark matter and dark energy.

Debate Nine

Scott Reeves vs. Ex Epsylon

Comments on YouTube video Gravitational Wave Hoax - LIGO fake blind injection discovery by Russ Brown

https://youtu.be/oed1Uqx9tQE

A quick note on the following. Ex Epsylon's comment came in at the last minute, as I was formatting this book. At that point, I had pretty much decided I had made my points, was tired of debating, and so had decided to "hang up my hat" for the time being. Thus, I did not post my response to Ex Epsylon on YouTube, since as I've said I had already decided to take a break from debating, and figured that posting a reply would only elicit further response, responses to which I would feel compelled to respond...and the debate would continue. So the following "debate" is very short, and my response can be found only in this book. This isn't any sort of a comment on Ex Epsylon; I would have welcomed debating him – if only he had chimed in earlier. Who knows? Maybe after a few months I will jump back into the fray, and further debate with Ex Epsylon will appear in a Volume 2.

Ex Epsylon wrote (in response to my earlier comments to MomoTheBellyDancer):

But of course Earth is the center of EARTH's observable universe +Scott Reeves, as from somewhere in the Sombrero galaxy the center of the observable universe will be the Sombrero galaxy, nobody disputes your Lapalissade.

But it has nothing to do with either the geometric or the gravitational center of the universe, that might or might not be in the same region, due to possible differences in mass distribution throughout the entire physical universe. But if we agree on the validity of the Big Bang / Big Bounce theory they should both reside in the vicinity of the point of origin of space-time.

You have quite an hypocrite attitude in this debate, you switch from hard science to points of semantics as it suits you best.

Try to maintain at least some intellectual honesty please, it's very difficult not to dismiss your assertions out of hand otherwise.

Scott Reeves wrote (but didn't post to YouTube):

"But of course Earth is the center of EARTH's observable universe +Scott Reeves, as from somewhere in the Sombrero galaxy the center of the observable universe will be the Sombrero galaxy, nobody disputes your Lapalissade."

That is what mainstream theories hypothesize. But that hypothesis has yet to be tested. Would an observer in the Sombrero galaxy be able to see 28 million light years beyond the edge of Earth's observable universe? Or is an observer in the Sombrero galaxy just 28 million light years closer to the actual edge of the entire universe, and thus observably NOT at the center of his/her/its own observable universe? Let's all travel to the Sombrero galaxy, set up a telescope and find out. In other words, let's do some actual science, instead of sitting on Earth and making unsubstantiated claims about what non-Earth-based observers will see when they look at the universe.

"But if we agree on the validity of the Big Bang / Big Bounce theory"

We don't. Unless the Big Bang theory can accommodate the observational evidence of Earth's central position at face value. Which the standard version can't, because it assumes, and depends upon, the validity of both the Copernican and the cosmological principles. Those are assumptions I'm not willing to make. The Big Bang theory as currently formulated depends upon the existence of a larger universe beyond Earth's observable universe (i.e. "from somewhere in the Sombrero galaxy the center of the observable universe will be the Sombrero galaxy"). Therefore, for the Big Bang to be a properly scientific theory, scientists from Earth need to go to the edge of Earth's observable universe and confirm that they can see a universe beyond. None of this unscientific attitude of, "Oh, we know what we would see if we did such a thing, so we don't need to do it."

"You have quite an hypocrite attitude in this debate, you switch from hard science to points of semantics as it suits you best."

Give an example of where I am arguing points of semantics, please.

And anyway, semantics are important. For example, if you say you disagree with THE geocentric model, you're saying something completely different than if say that you disagree with A geocentric model. Because there are actually two models: the relativistic geocentric model, and the absolute Geocentric model. If you say you disagree with the geocentric model, you're being very imprecise. Which geocentric model do you disagree with? You could be illustrating your ignorance of Relativity, or you could be taking an anti-Relativist position, in which case you're rejecting the relativistic geocentric model and advocating the absolute Geocentric model. So if that's the sort of thing you mean by your claim that I'm arguing semantics, then you're incorrect. I'm not arguing semantics. I'm exposing an imprecise choice of words on the part of my opponents that masks the fact that my opponents don't actually understand their own position on the subject, and shows their ignorance regarding what the true argument is. The argument isn't about the truth of geocentricity itself, but rather about exactly which sort of geocentric universe we live in

"Try to maintain at least some intellectual honesty please"

Since when does insistence upon strict adherence to the scientific method equate to lack of intellectual honesty? You've got it exactly backward. Anti-geocentrists, a group which should not include anyone who supports Relativity, but oddly enough, it does, are the ones who aren't being intellectually honest.

Appendix I

Geocentrists are anti-science.

Untrue. Geocentrists have very scientific and reasonable rebuttals for everything anti-geocentrists can throw at them. Geocentrists are very pro-science in their insistence that people who call themselves scientists should actually stick to science and accept the observational evidence at face value: Earth is at the center of the universe. Let's have none of this pseudo-scientific mumbo-jumbo about every point being at the center of its universe.

Modern technology wouldn't work if Earth were at the center of the universe.

Why is that? Exactly which technology wouldn't work if Earth were at the center of the universe? The prime one that is usually put forth in support of this foolish statement is GPS. But as shown previously in this book, the GPS corrections are predicted by absolute Geocentricity as well as Relativity, and both use exactly the same equations to correct the GPS clocks. So which other technology wouldn't work if Earth were at the center of the universe? Tell me. Which technology? The answer is: ALL of our technology would work if Earth were at the center of the universe. If some of our technology wouldn't work in an Earth-centered universe, then that means Earth is in a demonstrably inferior place in the universe, which is just as deadly to Relativity as Earth being in a superior place. So if even a single piece of our technology did not work in an Earth-centered universe, Relativity would be an invalid theory, which is okay by me, since that has been my contention all along. Either way, for me, geocentricity is a disproof of Relativity. Anyone making the statement that modern technology wouldn't work if Earth were at the center of the universe is actually taking a stand against Einstein. If people are happy taking such a stand, more power to them. Welcome to the club, here is your ID card.

The fact is that all our technology would work if Earth were at the center of the universe. There would just be a different dominant theory used to explain the science behind that technology.

And anyway, you'll often hear several versions of this argument. One says that modern technology wouldn't work if Relativity were invalid. Another says that modern technology wouldn't work if quantum mechanics were invalid. Yet another says that modern technology wouldn't work if Earth were at the center of the universe. The latter statement sort of rolls into the one about modern technology not working if Relativity were invalid, since if Earth is absolutely at the center of the universe, then Relativity is invalid. So basically we're left with the two statements about modern technology not working without quantum mechanics or Relativity. So which statement is true? Because, as physicist Brian Greene was quoted elsewhere in this book, "As they are currently formulated, general relativity and quantum mechanics cannot both be right" (The Elegant Universe, pg 3)

So depending upon which bit of technology you're claiming is based upon Relativity, and which bit is based upon quantum mechanics, some portion of our modern technology should not be working. And yet it does. So any statement that begins "Modern technology wouldn't work if..." is a fallacious statement, regardless of whatever follows the if.

Geocentrists are all religious kooks who only believe it based upon their need for God and a literal interpretation of the Bible.

This is demonstrably untrue. Myself being a prime example. Read back through this book and show where I presented a religious argument in favor of absolute Geocentricity. Geocentricity has plenty of non-religious arguments in its favor if one cares to look beyond his/her mockery and actually do in-depth research into the subject. I challenge anyone reading this book, who has what they deem to be a valid natural phenomenon that contradicts geocentricity, to actually search for the scientific, geocentric explanation for the phenomenon. Because I guarantee you it is out there, and presented in great detail. And it's most likely an absolute Geocentric explanation, because relativistic geocentrists seem to have dropped the ball on finding relativistic geocentric explanations for a whole host of phenomena. To hear all the people I debated with explain it, they have no workable geocentric explanations for earthquakes, satellites, the orbiting sun, Focault's pendulum, etc – all things for which relativists damn well better find geocentric explanations if they want to claim that Relativity is a valid theory. So get cracking, all you relativists reading this.

Also, it is no more true that all geocentrists are religious kooks than it is true that all scientists are atheists. And if all scientists did in fact happen to be atheists, it would no more constitute proof that their theories were correct than it would constitute a disproof of geocentricity if all its proponents were religious. Only the empirical support for a theory is decisive, not the number or philosophy of the theory's proponents. And geocentricity is at least as equally well supported by the empirical evidence as any other theory. If you disagree with that statement, you are both an anti-Relativist and an anti-geocentrist. And if you are correct in your disagreement, I win, because being an anti-Relativist is what leads me to absolute Geocentricity. If they're both incorrect, let's start looking for the correct theory of reality.

Appendix II

The following are a few things I wrote in anticipation of needing them for a response, but never actually used.

Fine. For the purposes of this discussion, I'll retract what I said about a geocentric frame being inertial, and allow that technically a geocentric reference frame is non-inertial due solely to the presence of gravity. But this doesn't change the fact that general relativity provides no answers as to whether the Earth is actually orbiting the sun and rotating. In such a geocentric frame, Earth is stationary and non-rotating even as it experiences a gravitational field. So Earth is non-inertial due to solely gravity, while the rest of the universe is non-inertial due to its rotation around the stationary Earth. There is no help for your claim that acceleration and gravity answer the question of which frame is truly in motion.

Consider this: from the viewpoint of an observer stationary within a sun-centric reference frame, the sun is stationary and non-rotating, while all the planets and the rest of the universe are accelerating around the sun, even as the sun-centered frame itself is non-inertial solely due to its own gravitational field. How do you then argue that the sun is actually not stationary and is itself orbiting the center of our galaxy? How do you then prove that our galaxy itself is not stationary and is actually orbiting the center of mass of the local group? Etc, etc, each time expanding the scope of the system of masses under consideration?

There is no objectively decisive factor as to which body is actually in motion. Inertial vs. non-inertial is never a determining factor, regardless of whether the gravitational field in question is "real" or "fictitious." The decisive factor is a baseless assumption that the reference frame in question is in motion relative to something else.

Until, of course you've reached the point where you're considering all the mass in the universe as a whole. Once you've pulled your perspective back that far, it becomes apparent which objects are actually in motion relative to the whole.

Fine. Then the heliocentric reference frame is non-inertial as well, so it is definitely moving. As is the Milky-Way-centric reference frame. As is the Local-Group-centric reference frame. As is the Virgo-Supercluster-centric frame. As is the Laniakea Supercluster. As is whatever lies at the next larger scale. And the next. And the next. Or perhaps not. Perhaps one of these levels is stationary relative to something else. But then you're just assuming that it's that level that is stationary rather than the something else.

But you're eventually going to reach a scale where you have encompassed all the matter in the entire universe. And that entirety cannot possibly be in motion relative to anything else, because there is nothing else to be in motion relative to. Which means that all the matter in the universe as a collective whole is absolutely stationary. Which means then that you can work your way inward and determine which objects are actually in motion relative to the whole. Those things that are in motion relative to the whole are therefore in absolute motion, meaning that relativity is an invalid theory, meaning that there is an absolute reference frame.

Fine. For the purposes of this discussion, I'll retract what I said about a geocentric frame being inertial, and allow that technically a geocentric reference frame is non-inertial due solely to the presence of gravity. But from Earth's perspective, the rest of the universe is rotating around the Earth and is therefore also a non-inertial reference frame, due solely to its rotation. And the heliocentric reference frame has the same problem.

Look, if you relativists think earthquakes disqualify the geocentric model, then fine. There's a geocentric reference frame, but it has inferior explanatory power. That means Relativity is an invalid theory, which is what I was saying long before I began espousing absolute Geocentrism. I began advocating absolute Geocentrism because I thought that it was the only alternative to Relativity. But since you have invalidated Relativity by determining that the geocentric reference frame is an inferior reference frame, that means there is an alternative to both Relativity and absolute Geocentrism, and that's fine with me. Let's do some science and figure out what that alternative is. We know that it is non-relativistic and non-geocentric. That's our starting point.

Appendix III

Is my contention that scientists say Earth is at the center of the observable universe correct? Of course it is; I don't make baseless claims. Here are numerous quotes on the subject.

From Wikipedia's entry on "Earth's location in the Universe": Since there is believed to be no 'center' or 'edge' of the Universe, there is no particular reference point with which to plot the overall location of the Earth in the universe.[8] Because the observable universe is defined as that region of the Universe visible to terrestrial observers, Earth is, by definition, the center of the observable universe.[9]"

Note the phrase "Since there is believed to be…" It's merely a *belief* that Earth is not at the center of the universe as a whole. It's not something that has been empirically proven. If it were empirically proven, I doubt the word "believed" would have been used. But then, this is Wikipedia, after all, and nothing on Wikipedia is reliable. And yes, of course I'm being sarcastic.

From the Wikipedia entry on "Observable Universe": "The **observable universe** consists of the galaxies and other matter that can, in principle, be observed fromEarth at the present time because light and other signals from these objects have had time to reach Earth since the beginning of the cosmological expansion.

Assuming the universe is isotropic, the distance to the edge of the observable universe is roughly the same in every direction. That is, the observable universe is a spherical volume (a ball) centered on the observer. Every location in the Universe has its own observable universe, which may or may not overlap with the one centered on Earth."

From http://www.universetoday.com/109462/where-is-earth-located: "What a strange coincidence for you and I to be located right here. Dead center. Smack dab right in the middle of the Universe. Certainly makes us sound important doesn't it? But considering that every other spot in the Universe is also located at the center of the universe...Every single spot that you can imagine inside the Universe is also the center of the Universe."

From http://www.space.com/24073-how-big-is-the-universe.html: "Like a ship in the empty ocean, astronomers on Earth can turn their telescopes to peer 13.8 billion light-years in every direction, which puts Earth inside of an observable sphere with a radius of 13.8 billion light-years. The word "observable" is key; the sphere limits what scientists can see but not what is there."

Note the careful avoidance of the word "center." "...which puts Earth inside of an observable sphere..." should more aptly read "...which puts Earth at the center of an observable sphere..." Also note the last sentence about "observable" being key in that it limits what scientists can see but not what is there. "Observable" is also key to the scientific method, so as far as the scientific method is concerned, contrary to the article, the sphere limits both what scientists can see AND what is there. If science can't observe something, that something ain't there as far as science is concerned. Sorry, all you Relativists and Big Bangers, but your theories depend on the existence of something that cannot be scientifically demonstrated to exist. Hence, your theories are pseudo-scientific.

I could go on quoting, but I'll leave it to the reader look more deeply into the matter for him/herself.

Just remember, any time you hear someone say, "EVERY point will look like it's the center," which is the Copernican Principle, that de facto means that Earth will most definitely look like it's the center. But as I have said repeatedly throughout this book, the claim that every point will look like it's the center is an untested hypothesis.

I'll leave you with this thought: why do we need either a Copernican or a Cosmological Principle if astronomical observations didn't show Earth to be at the center of the universe? And yes, I did say at the center of the universe. That whole "at the center of the OBSERVABLE universe" thing is an artefact of the Copernican Principle. According to astronomical observations, Earth most definitely is at the center of the universe, and the cold, hard, unacceptable fact of those observations is what drives modern scientists put forth the Copernican hypothesis.

Made in the USA
Middletown, DE
26 February 2025

71929988R00125